W9-AXK-108

Being a Dog

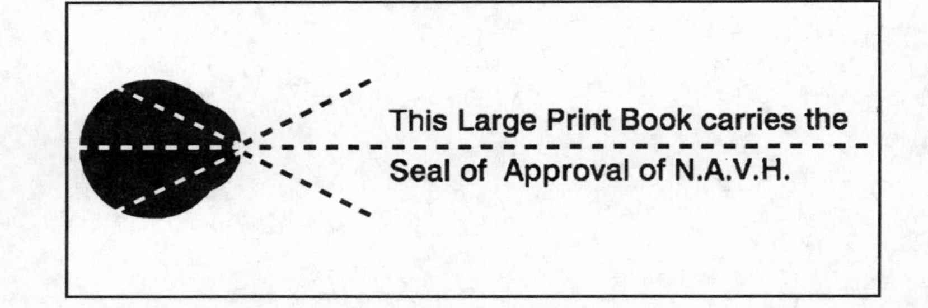
This Large Print Book carries the
Seal of Approval of N.A.V.H.

Being a Dog

FOLLOWING THE DOG INTO A WORLD OF SMELL

ALEXANDRA HOROWITZ

THORNDIKE PRESS
A part of Gale, Cengage Learning

Farmington Hills, Mich • San Francisco • New York • Waterville, Maine
Meriden, Conn • Mason, Ohio • Chicago

GALE
CENGAGE Learning®

Thorndike Press® Large Print Popular and Narrative Nonfiction.
The text of this Large Print edition is unabridged.
Other aspects of the book may vary from the original edition.
Set in 16 pt. Plantin.

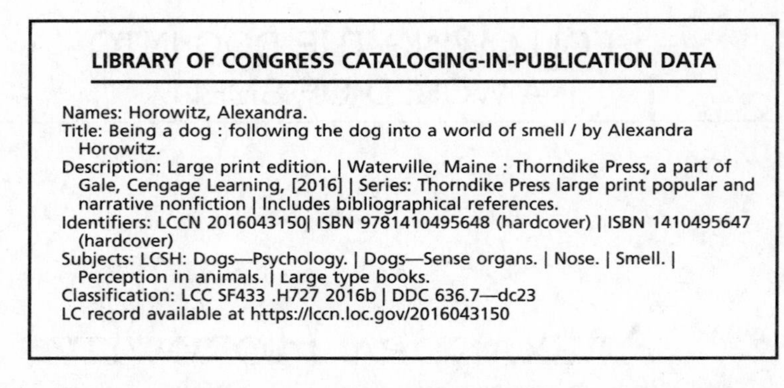

LIBRARY OF CONGRESS CATALOGING-IN-PUBLICATION DATA

Names: Horowitz, Alexandra.
Title: Being a dog : following the dog into a world of smell / by Alexandra Horowitz.
Description: Large print edition. | Waterville, Maine : Thorndike Press, a part of Gale, Cengage Learning, [2016] | Series: Thorndike Press large print popular and narrative nonfiction | Includes bibliographical references.
Identifiers: LCCN 2016043150| ISBN 9781410495648 (hardcover) | ISBN 1410495647 (hardcover)
Subjects: LCSH: Dogs—Psychology. | Dogs—Sense organs. | Nose. | Smell. | Perception in animals. | Large type books.
Classification: LCC SF433 .H727 2016b | DDC 636.7—dc23
LC record available at https://lccn.loc.gov/2016043150

Published in 2017 by arrangement with Scribner, an imprint of Simon & Schuster, Inc.

Printed in Mexico
1 2 3 4 5 6 7 21 20 19 18 17

for my Dad

CONTENTS

1: NOSE OF A DOG

Finnegan's is ebony-black, moist, and dappled, two cavernous bass clefs at its front. Upton's is cleft with a visible valley, the whole thing guarded by short whiskers standing at attention.

These are my dogs, and these are my dogs' noses.

Before I became a research scientist studying dog cognition, I didn't think much about the dog nose. It may have been frowned at when poking impolitely at the privates of visitors to my home, or dabbed with peanut butter to encourage the swallowing of a pill. Even then, though, little regard was given to the nose itself — to its form, to its movement, to the impossibly convoluted and complex vault it opens into.

This oversight isn't restricted to dog noses. We rarely study the nose on each other's face. At the nose prominently displayed — indeed, jutting out, leading the way for the rest of the body. Without looking, try to describe your partner's nose, or your mother's. If it's not beakish or buttony, well, *it's just a nose.* Two nostrils dangling from a squashed, fleshy tetrahedron.

I gaze at my son's nose, but mostly at its surface — where the freckles have begun to assemble on his fair skin. But the snout of the dog gets my full attention. Now I look at dogs nose-first. For I am besotted with dogs, and to know a dog is to be interested in what it's like to be a dog. And that all begins with the nose.

What the dog sees and knows comes through his nose, and the information that every dog — the tracking dog, of course, but also the dog lying next to you, snoring, on the couch — has about the world based on smell is unthinkably rich. It is rich in a way we humans once knew about, once acted on, but have since neglected.

By smelling, tapping into this sensory resource that we have but that we largely ignore, the dog has become an informant. Working dogs, trained to tell us what they naturally know, detect the presence of illegal substances and unwanted pests. But the dog also knows about the upcoming weather, the way afternoon smells, and whether you are sick or upset. Every inhaled gulp of air is full of information. It holds the odors of people who have recently passed by, leaving olfactory trails in their wake. It catches pollens and plant notes carried on the breeze. Each noseful captures the traces of animals who have walked, run, cowered, eaten, or died nearby. It traps the electric charge and round humid molecules from distant rainstorms.

This book is an exploration of what the dog's nose knows as has never been done before. What does your dog smell on you, on the ground, or with his nose deep in the fur of another dog? What does he know about you — that you might not know yourself? What is it like to smell the world with that amazing nose leading the dog through his days?

To find out, I tracked the tracking dogs. Over the last years I have watched detection dogs grow up, be trained, and find their quarry, be it drugs, food, or people. With my Dog Cognition Lab at Barnard College, I have researched the pet dog's experience of himself, other dogs, and the smells of the human world in which he lives. I have spoken to scientists who study and model the dog nose, and to trainers and handlers who follow it. It is an examination of all aspects of the dog's olfactory world, and the magnificent organ that leads to it.

But it is also an exploration of the noses on *our* faces. We have untrained ourselves, we humans — unlearned how to smell, over millennia. We are out of the habit of smelling. Why, you may not even have smelled this book yet, though it is but inches from your face. I find people who *are* smelling, and practice their methods.

Having myself committed a lifetime of not smelling, I let my dogs' behavior counsel me: I have ventured to be a little more like a dog myself. In my book *Inside of a Dog* I began an imaginative leap into what it might be like to be a dog — and here I leap with four feet. I try to take my nose to the places the dog nose goes. And I sniff.

I begin this process by learning more

about our own sense of smell. And then I train my nose to better conjure what it might be like to have the mind and nose of a dog.

My inspiration and guides are our family's own dogs, Finnegan and Upton. Both are highly charismatic mutts. My husband and I met Finnegan's nose through the bars of a crate at a shelter that imports unwanted dogs from the South. He was four months old, had had ringworm and parvovirus, and, though recovering, was skinny and a bit sickly. I should say that I don't go to shelters much, because when I do, I inevitably walk out with a new animal. When I first saw him, in that crate at eye level, he wagged mightily, accepted a finger poked through the bars for licking, poked his nose through reciprocally, and then, as we moved on, sat down patiently. I looked back at him often: he was sitting and . . . waiting. We took him out of the crate to meet him, and he moved between my husband and me, looking at each of us in the face. Weary, he leaned against me. That was that. We took him home.

Finn is now eight years older. He still has the look of that puppy who leaned. While his coat is a glossy black, as though we polish him daily, it is his way of *looking* that most captures him. One gets the feeling that he is always aware of what is happening. His eyes penetrate us. They follow us, they check with us when another animal is misbehaving, they look to us dolefully when we head out the door. Ears back, eyes full, he is hard to leave. He does not look simply with his eyes: his nose examines us. When we return home, he sniffs us, as close as we will let him, in an exploration of where we've gone, what we've eaten, whom we've touched or pet. Never have I returned home, having met a dog on the street, and escaped Finn's examination.

I tend to think of Finn as a "professional dog." He is eminently civilized: without our explicitly teaching him, he fell into line with what we expect from a dog in the house. He swallowed the little culture of our family whole. Upton, by contrast, whom we met three years into his life, is feral by comparison. He had been relinquished to the shelter from which he was initially adopted three years prior. We've seen his first photos: a small-bodied hound, ears too large for his head. Nose a blur. Well, his head and body grew up: he's a large, brindle, hound mix, with full eyes and a corkscrew tail. Whiskers punctuate his snout; his jowls droop. He is a dog's dog, incurably friendly with any dog, and has a gangly, goofy run. There is no photo of Upton in which he looks streamlined, athletic, or svelte. In running, his jowls flap, he tilts, his ears go akimbo. He is generously silly. Alas, he was also not a city dog before we knew him, and he is easily startled by every and any sound — car door, garbage truck, garage door opening, street sign shaking in the wind, jackhammer, a bag flapping in a tree, a person suddenly appearing from around the corner . . . you name it. For this reason, while I have followed both of their noses to whatever sundry and sordid spots they're smelling, it is only Finn who I bring out to do new investigative smelling work. He could half write this book himself.

Take a breath (through the nose, please). We are going on a journey through scents and smelling, with a tour through the improbable science about the olfactory abilities of the dog — and the abilities of our own noses, waiting for us to discover them. By following the dog's lead, we can learn from him about what we are missing — some of which is beyond our ability to sense, and some of which we simply need a guide to see. The world abounds with aromas, but we are spectacle-less. The dog can serve as our spectacles.

In so doing, we may also see how to return to that perhaps more primal, so-called animal state of knowledge about ourselves and the world that we have forgotten in a culture wrought of technology and lab tests. To follow animals is to become more attuned to our own existence. To follow *dogs* is to begin to apprehend the experience of our silent, loyal partners through our days.

2: Smeller

> The smell of that buttered toast simply talked to Toad, and with no uncertain voice; talked of warm kitchens, of breakfasts on bright frosty mornings, of cosy parlour firesides on winter evenings, when one's ramble was over and slippered feet were propped on the fender; of the purring of contented cats, and the twitter of sleepy canaries.
>
> — Kenneth Grahame, *The Wind in the Willows*

In my home, the library is rife with reminders of how little we might understand what an animal smells. Not my library — my six-year-old son's. Some of the most fabulous writers for children were highly attuned to smell. Roald Dahl's child-smelling monster and sweet-smelling chocolate palaces are matched by the animal protagonists from William Steig's imagination. His character

Dominic, a nomadic, adventurous, and canine soul who leaves his barnyard-animal friends to explore the world, gives them an apt good-bye: "I embrace you all and sniff you with love." Dominic's "all-knowing" nose directs him in his travels. He sniffs out an evil fox — whom he "would have recognized [by] smell merely by sniffing something he had touched only lightly a year earlier." He smells out tea, sugar, and milk for a snack. His nose discovers a sick pig, a witch-alligator, the residents of an unknown town. "Dominic always noticed smells first," Steig wrote. All of the animal residents of Kenneth Grahame's *Willows* notice the "warm, rich, and varied smells — that twined and twisted and wreathed themselves at last into one complete, voluptuous, perfect smell that seemed like the very soul of Nature taking form and appearing to her children."

Even that is not the half of it.

I live with two quadrupedal reminders of the keenness of the dog nose in my house. We are agog when they find minute particles of food that long ago skittered off a young child's plate. But clearly this is not the extent of their olfactory acuity; it is only the extent of my daily awareness of it.

Scientific measures of the sensitivity of

the dog nose are limited more by the capacity of measuring devices — and dogs' interest in submitting to them — than by their noses. Both pet dogs and tracking dogs have been put through their paces in a variety of threshold detection tasks that ask how diluted an odorant can be before a dog stops noticing that it's there. Pick out a canister with the smell of banana, amyl acetate, from a series of non-banana-smelling canisters, for instance. Dogs keep finding the banana until it is diluted to 1–2 parts per *trillion:* a couple drops of amyl acetate, one trillion drops of water. Early research with one very cooperative fox terrier concluded that she could essentially detect one milligram of butyric acid — think smelly socks — among 100 million cubic meters of air. You'll notice your spouse's smelly socks the moment after they are removed in the bedroom: that's around forty cubic meters of very socky-smelling air. The dog knows if someone's removed his socks in a room bigger than the gargantuan vehicle assembly building at NASA's Kennedy Space Center in Florida, made to put the space shuttles together. Any dog in the nearly four million cubic meters of Space Center would be alert to sweaty astronauts.

Explosives-detection dogs smell as little as a picogram — a trillionth of a gram — of TNT or other explosive. What might it be like to notice a picogram of an odor? Since explosives dogs come to have very fond associations with their search odors, let's think of an aroma pleasing to our noses: cinnamon rolls cooking in a home kitchen. The average cinnamon roll has about a gram of cinnamon in it. Sure, the human nose is on it, from the moment we open the door of the house. Now imagine the smell of one trillion cinnamon rolls. That's what the dog coming in with us smells when we enter.

The sensitivity of the dog's nose can also be gauged by simply looking at his behavior. Hunting and tracking dogs naturally follow scent trails of prey or people who have passed before — in some cases, several days before, over difficult terrain. Watch video of a Discovery channel show's host trying to outwit a bloodhound by crossing a river, spraying himself with deodorant, laying down sausages as a diversion, then doubling back and going a different way. Then watch the dog follow the path he had run, crossing the river, noticing (but forsaking) the sausages, and then doubling back and easily catching himself a TV personality.

Encountering a trail in its middle, a tracking dog needs to sniff only *five* footprints — laid in less than two seconds — to tell which way the person who made those prints was going. Each footprint holds a certain amount of that person's odor, and as the intensity increases between step one and five, the dog has his answer. Run other people over the same path, intersect with other trails; the dog can still find his man.

Dogs are so good at finding their man that in the Netherlands, Germany, Poland, and a handful of other countries, courts have sanctioned evidence from dogs presented with scent-identification lineups. The human-scent lineup is *not* what it sounds like: a dog sniffing his way down a line of suspects and innocent people (called fillers), stopping at each, sniffing, and assessing.* Instead, the dog sniffs his way down an array of metal bars that have been handled by the suspect and the fillers. Dogs pick out the smell that matches an odor

* Although, how terrifying that might be for the suspect! And how potentially effective: surely a suspect would release those odorant molecules from each and every sweat gland available. Unfortunately, so might any simply anxious fillers waiting for their turn as sniffee and wondering what secrets this seemingly miraculous dog might have.

from a crime scene — thus picking out the criminal.

Your own dog, beside you right now, performs surprising and sometimes alarming feats of olfactory perception every day. Many of these behaviors are familiar to us; what is not familiar is the smell behind it.

As a first step toward discovering the acuity of the dog's nose, then, let us look at what a dog can and does smell in the course of an ordinary day. Our dogs exist in parallel with us, at our feet and by our sides, living in lockstep. Though we look them in the eyes and see how they *look* — gazing at us, peering off toward a distant barker — most of their behavior is about their noses, smelling the world.

Nice to Sniff You

Left to their own devices, most dogs will not pursue the dissipating trail of a stranger, or sniff metal bars one after the other. Dogs like sniffing other dogs. People like looking at people — and left in a room by themselves will look at images, static or moving, of people. Dog pinups are not popular among the dogs I know, but were I able to bottle the odor of the lean black-and-white pup down the street, it might make for a good diversion when our dogs are alone and bored.

Every owner has surely observed this. But do not think that the mutual sniff is meaningless, a mere sneeze. When one dog sneezes, his companion does not. But when two dogs meet, they sniff — and let themselves be sniffed — as a real communication. There is likely a pleasure in sniffing other dogs, but what we cannot see is the information gleaned. There is a measured flow to their smelling: both sniff each other at once or take turns politely, burrowing noses into fur. The fur bears odors from skin glands located in sniffable locations. And these odors are the key: they hold news of the dog from which they effuse.

Watching male and female dogs sniff one another, researchers found that males like to go for what they call "the tail area" (read: rump) first. Skin glands circle the anus, secreting odor. At either side of the anus — at "four and eight o'clock," one writer helpfully guides the clock-readers among us — are anal sacs producing a strong odor of *dog.* More specifically, they seem to convey *stressed dog.* When a dog is anxious, the sacs produce a skunky smell. Secretions from the sacs also serve as a topping to every poo. For this reason, some scientists consider this odor likely to be the dog's "signature" odor — his identification tag, written for smellers. Forty years ago, Dr. George Preti of the Monell Chemical Senses Center and his colleagues squeezed out the contents of some (semi) cooperative beagles' anal sacs. "I was a pioneer!" he told me. "With no followers." What they found after examining the constituents was that, though the smell might seem similar — and noxious — to most human noses, the samples varied greatly: enough to be a marker of each individual. Because dog researchers will do just about anything (apparently) in search of knowledge about their subjects, we now know that, even for humans, there are noticeable differences between individuals' secretions — ranging from a "somewhat neutral or slightly pleasant, doglike odor" on the one hand to a "sharp acrid" odor on the other. Thank you, scientists, for sniffing that.

Canids have distinct tail glands as well, at the base of the tail. You can locate them by either watching a male sniff a new dog or by looking for an oily patch of hair where the tail meets the back — so oiled from glandular secretions. In the fox, this supra-caudal gland produces odors notable even to our noses: a hint of violet in the red fox, muskiness in the gray fox. Insofar as these glands reflect differences in the levels of sex hormone, part of what those male dogs are keen on finding out about a new dog is who she is, sure, but also whether she is ready to mate.

Female dogs more often sniff faces first. Wolves, too, are especially keen on studying the smell of each other's head and muzzle. You may have a dog who noses into your ears or pokes around your nose and eyes. He is treating you like a proper dog: there are plenty of apocrine and sebaceous glands and secretions in dogs' ears, and eccrine glands on their noses. Perhaps this examination is less about mating and more about determining health and diet. Saliva smells wherever saliva gets, which in dogs is plentifully spread around the face and snout.

If you get close enough, you will smell the unique odor of your dog. What your dog smells like to you is probably due to the secretions of the apocrine glands, which are over his entire body. There is also an individual smell on the bottom of his foot pads. If you have not ever sniffed your dog's feet, for goodness' sake, now's that time. (You can be sure your dog knows the smell of *your* feet.) On dogs, glands grace the pads themselves and hide between the toes. They emit an odor distinctive enough — to other dogs, at any rate — that it might explain one of the all-time most puzzling dog behaviors: scratching the ground after peeing or pooing. I've known dogs who in moments of great excitement of any sort — a fine-smelling dog passes by; they have just finished a rousing, frenetic run — furiously scratch long, deep rivulets in the ground, as though rendering an exclamation point on the scene. If every scratch releases drops of odor, then this behavior probably serves as a signpost to lead other dogs to where they might find the mother lode of smells: the excreta or urine the scratcher left behind.

A sniff of another dog reveals who he is, but also information that only the more undignified gaze would catch as it falls. Dogs do not glance under the belly to see if the new dog is male or female. They *smell* male or female. Moreover, they may smell ready to mate, of being recently ill, or of having recently eaten. They smell their age: age is but a metabolic process, chemistry. And chemistry smells. Dogs smell of having had a bath or not having had one; they smell of having recently peed or very much having to pee. They smell of themselves, their status, and, likely, whether they are scared or happy or anxious.

MESSAGES LEFT BEHIND

Even without an actual dog to nose on her walk, your dog is not bereft. Happily, dogs have left behind an embarrassment of scented calling cards. Every footstep on a porous surface leaves odor; the clump of fur an owner has brushed off has oils and secretions from hair follicles. We carry home with us the smells of dogs whose owners we have visited; of melancholy dogs leashed outside stores who have let us tickle their ears; of indiscriminately friendly puppies met on the sidewalk who tumbled, slobbered, and dribbled all over us.

And then there is the pee. Anyone who has spent hours on hours on grassy knolls or lawns popular with dogs has witnessed the tragic Person Mark. A dog owner, letting down her guard, perhaps feeling a little weary, sits on the grass as dogs gambol about. Suddenly, without warning, a single dog tears away from the games, approaches the person's side or back, raises a leg and . . . urinates.

The person has been "marked." Should she stay seated, it would not be long before she would be counter-marked by another dog. She does not stay seated. She rises and people commiserate and scold the dogs between laughs. But the dogs do not see this as bad behavior, of course. This behavior has long standing. Dogs do it, bees do it, even hippopotami do it.

Scent-marking is the leaving of urine or other bodily secretions on a stone, stump, bush, or other protruding feature of the landscape: fire hydrants on city streets, tractor wheels in farming climes. The marked item becomes a scent post: an olfactory flag holding information about the marker, ready for the sniffing.

Classically, scent-marking is thought to be territorial. Classically, it has been — in most animals. And the means and placement are wonderfully more complicated than human flag-leaving. Muskrats leave an oily scent on a blade of grass; beavers leave a distinctive yellowish oil, castoreum, right on top of a pile of pond muck that they've gathered on the shore. Otters do them one better, making a whole scent-region by the water's edge by rolling downhill and defecating to top it off. Snowshoe hares even each-other mark: when courting, one leaps over the other, balletically, while spraying his partner. Dik-diks paw a communal dung heap, then scent-mark their path with their footprints; spotted hyenas leave anal-gland scents and scents from between their toes and use communal latrines at the edge of their hyena town. A cat "bunts" a post with his face, spreading odor from glands on his chin and cheeks. Both the hippopotamus and rhinoceros deliver a forceful urine mark; the rhino projectile-pees into a bush he's just torn to pieces with his horn, for good measure. The badger "bum-presses" a scent on the ground; the mongoose and female bush dog have been known to do handstands to spread their urine or anal-gland scents.

When scent posts are at the edge of social animals' territories — areas often patrolled and defended — it seems apt to call them territorial. By fencing your property, you are indicating that anyone who steps over the fence is trespassing. No other signage needed. But many posts are not left on an animal's territory at all: they are left as one walks a new area, or in a shared social space. Marks on what are called runways, or shared paths or trails, are not about claiming the path. In these cases the mark is probably conveying social information about who left it and the kind of animal he is.

After that first marking, some species *counter*-mark: peeing or rubbing over the mark another animal has left. While *eau de spotted hyena* might carry well on the breeze, most animals will bother to go right up to the fence-mark and examine it closely first. Any counter-marks may be seen as a challenge to the territory-holder, or a kind of response to the social-information leaver: *Here I am, too.* We know that counter-marking is not simply territorial, because it is not regularly followed by territorial challenges or relinquishments. It is also socially competitive: among house mice, the top-marking counter-marker is often the most popular house mouse.

So what about with dogs? Both wild and domestic dogs conspicuously mark and counter-mark objects. This is typically achieved through a sometimes gymnastic "raised leg display" (what scientists, in a fit of unneeded acronymizing, call the RLD), balancing on three legs while raising the fourth high. Done by both males and females, the RLD allows for a directional flow of urine, aimed to land (whether it does or not) on a vertical or semivertical object. Note that marking is not just *peeing* willy-nilly: urine-marking is the leaving of just a spray here or a mist there.*

*Physics researchers have found consistency among bladder-emptying peeing, which they dubbed the Law of Urination: on average, all animals, from dog to vole to elephant, take twenty-

Here's a surprise: in contrast to these other marking animals, domestic dogs do not mark territorially. Yes, you read that right. Dogs are not "marking their territory." How do we know this? Simply by looking at where dogs do — and do not — pee. Owned dogs do not mark the periphery of their homes. The apartment-living dog does not pee along the walls and threshold. (Does yours? That's another topic . . .) When living in fenced suburban yards, dogs do not assiduously line the border of the property with pee. Research in India on the massive free-ranging dog population there — stray dogs who actually might have home territories at risk of being wandered into by others — found that they, too, rarely mark on the boundary of a territory. Dogs walked along shared paths and parks could not verily consider these areas their "territories," given the occasionalness of their occupying them — and indeed they show no accompanying behaviors that would indicate that dogs feel that a path is "theirs."

Instead, dogs are great runway markers. Consider where your dog pees: on the lamppost along a shared path, the one tiny bush

---

one seconds to pee everything out. Marking, by contrast, takes one second or three.

along the country road, the garbage pail at a driveway's end that wasn't there yesterday. Notably, they spend a lot of time sniffing all the possible marking sites, but they do not counter-mark *all* of them. A sniff may be followed by a quick look around, a scratching at the ground, or even a chattering of the teeth — part of smelling the hormones in the urine.

So what are dogs saying to each other as they load a fireplug with layers of pee? Most likely, it is social information being left. Calling it pee-mail is not far off. They are telling each other who they are, and, intentionally or not, revealing a host of other information: their sex, whether they are a female in estrus, what they ate, how they're feeling, their health. The few studies asking how and when dogs mark found that non-neutered male dogs marked more, counter-marked more, and teeth-chattered more than their neutered brothers or than females. But all mark and counter-mark (although some "adjacent" mark, wildly missing the target, intentionally or not). Simply the amount of time that dogs, left to their own devices, spend sniffing indicates that the mark has reams of information.

But the dog scent-mark is the perfect graffiti: you must have a secret nose key to unlock its specific message. Thus far, human researchers have failed to decode the mark. Part of the reason for this is that we are not *asking the dogs:* while few animals answer questions about a behavior in easy-to-parse sentences, often what he does after the behavior serves as an answer. If a firefly flashes his light three times and a dozen female fireflies come hurtling toward him to mate, we get a pretty good idea what "three flashes" roughly means.

This is the reason that I applied to the NYC Parks Department to do research in their parks. While it would cost nothing, disturb no habitat or organism, and be unobtrusive, the proposal was, well, unusual. I proposed to put a "pee post" in one of their parks, Riverside Park, and watch what happened. How many dogs would sniff a post that's been peed on? How often do they over-mark? How high and accurate are they in their aim? Do they come back to see their own handiwork? And what do they do after sniffing and peeing?

Six weeks later I learned that my proposal had been accepted. On a London plane tree I mounted and secured a camouflaged motion-activated camera. I aimed it toward a short post — about the size of a standard metal fencepost — which was conspicuously placed by a popular dog-walking path.

For a week at a time the post brought curious noses, and the camera witnessed it all. What it captured was dogs taking in the information in marks — and leaving marks for others — but rarely following through. Dogs who sniffed marks along the path often then looked back and forth on the trail for the mark-leaver. When the other dog was still nearby, they showed signs of wanting to pursue him. But, as a species whose walking route is usually determined by a person, not by themselves, they were usually prevented from pursuing the good-smelling dog by a leash. Counter-marking was surprisingly rare: sniffings outpaced markings. When a dog marked, it seemed to take the place of actual interaction: *If I can't sniff you, I'll leave my card for you here.* But even off-leash dogs never returned to see if their messages had been scribbled over. As for what the dogs read in the bulletin board of marks that quickly dotted the post, that remains a mystery. Without a territory to guard, they still post tiny, odorous flags. But they never check who is saluting.

ROLLING IN IT

As a scent-mark gets the animal's scent on the ground, scent-rolling or -rubbing gets the ground's scent on the animal. Or, rather, the scent of whatever-it-is that they are rolling in. Dogs are unrepentant scent-rollers — and the scents they roll in are often extremely stinky. In this they share their tastes with other mammals: a list of common desirables includes "meat (fresh or decaying), vomit, intestinal contents, cheese, engine oil, perfume, insecticide, and the feces of other species." Other researchers have added "raisins, beetles . . . cigarette butts, hard candy, human bed pillows, and many spots with nothing detectable to a human nose or eye" to the list of rollables.

Commonly, a dog sniffs at the scent, getting right down into it — although in the case of decay or discharge this seems a bit unnecessary — and then hurtling his head or shoulder into the source, followed by his neck and back, often squirming supine with apparent glee. I am reminded of nothing more than the cat's enrapture with catnip. In both cases it seems to spasm the neural circuitry responsible for play, sex, and eating all at once. Thus both species may roll ecstatically, bite and paw at the scent, and rub their faces in it.

Why the dog does this is the source of uncertainty by scientists and owners alike. Theories come in a few flavors. One is the "camouflage" theory: by matching their own bodies to the environmental odors, they are more likely to be seen as territory holders. African wild dog females roll in the urine of the males whose packs they then attempt to join: they may be more likely to be accepted if they smell like home. Another is the "popularity" theory: it might enhance their social standing to be covered with something as clearly desirable as a very, very stinky stink. Spotted hyenas who were scented with carrion on their shoulders got more grooming attention by their pack than those who were dotted with camphor. Finally, the hedonic theory: it is simply pleasurable. Perhaps by festooning themselves with the perfume of a decaying animal, the fragrance can be enjoyed later. New sources of smell are particularly interesting: should you want your dog to smell like your perfume, try pouring a little on the grass. Next to the days-old cat poo.

These are but the most ordinary examples of dogs' use of smell. But then there are the seemingly extraordinary dogs, the detection dogs, weaned off pee-marking and now smelling professionally. They smell things we not only cannot see, but often things we cannot imagine.

Detection dogs have been trained to find just about anything, from the lock to the stock and barrel. Dogs find, we know, explosives, accelerants, and land mines. They find missing people, still alive — and cadavers, on land or underwater. They can smell out drugs and counterfeit goods. But they can also detect illicit cell phones in prison and imported sharks' fins in suitcases; termites, fire ants, and the red palm weevil, which kills palm trees used both ornamentally and for their dates; screwworm, nematodes, and bedbugs; invasive knapweed in Montana and invasive brown tree snakes in Guam; the hard-to-spot Northern right whale in the sea and the Amur tiger on land; the feces of black bears, fishers, bobcats, maned wolves, bush dogs, and turtles; birds killed on wind farms and dairy cows who have come into estrus. Provided that it has an odor, it can be smelled out by a dog. There are now dogs put into service to find other, lost dogs.

As we will see, these are extraordinary acts, but these are not necessarily extraordinary dogs. Any dog can perform surprising feats of discovery and identification. But only some dogs have learned to bother to tell us if they notice where a missing person has got to, or if a traveler is bringing a single guava across the border. Those are the ones trained in persistence and communication with their handlers, but they all have the same quality nose. Dogs seek out smells, they roll in smells, and if we ask them to, they act on smells.

This has made me realize what else in my dogs' lives has an existence through smell. Things that smell before they appear, things that do not appear to have a smell. Things that we might think *don't* smell (us) and things that don't smell as we think (them). Things that start to draw a different picture of what it might be like to see the world nose-first.

Smell of Us

One of the cheeriest genres of YouTube video compilations is that of the short, gleeful scenes of dogs greeting soldiers returning home. Whether the owner's deployment was long or short, the dogs erupt into that singular, irrepressible doggy celebration: bounding, tails wagging maniacally, whimpering, rolling and wiggling on their backs, grinning, spastically weaving under the soldiers' legs or between their arms — sometimes all at once. There is little doubt that the dogs remembered, loved, and missed their people.

In some of the videos, though, this recognition is at first in doubt: as the (often uniformed) person arrives or enters the home, the dog barks, approaching guardedly, tail down and ears back. He does not know this person. But then comes a magical moment of transformation. Look closely at the videos . . . pause on the frames when the dog's allegiance is in doubt. Watch his nose. Each dog lifts his nose in the air, catching a whiff on the breeze. Or sniffs first, one offered hand, then the other. In an instant, the stranger is transformed — into the person the dog knows and longed to see.

To our dogs, each of us is encircled by a cloud of scent that is as familiar to them as our image in the mirror is to us. We *are* our scent — and it is not the smell of our shampoo. Your dog would have no trouble picking you out from a line of people, even if you didn't dissolve into cooing, bend down, and tickle his head. Instead, he could pick out the particular human bouquet of *you* that is a mix of oleic, palmitic, and stearic acids; trained dogs notice if any of the ingredients in the mix is a few micrograms more or less present.

You may see a glimpse of evidence of this in your dog's anticipation or seeming foreknowledge of when you are coming home. Because such a perception is foreign to us, some label psychic anything the dog foresees — the hour of our return home, the earthquake that we notice only with the shakes. But with our smells and sounds, we announce ourselves with olfactory cowbells and the noisiness of a skunk's spray. To the dog, we arrive before we do and stay after we leave.

Many people have pointed to dogs who seem to "know" when their owners are returning as a special kind of skill. I think, rather, it is the special skill of *smell.* A few years ago I concocted a thought experiment to test just how much smell mattered to a dog in sensing when his owner was returning home. Rather than assume that the dog was smelling or hearing the owner through the door, I suggested that there was a potent combination of two forces leading to these dogs' abilities. The first is the distinctness of our smell to our dogs. The second is the ease with which dogs learn our habits: we leave and return reliably — if not always at precisely the same time. So, how can your pup know when you are returning from work, given that the sun sets at a different time each day? Well, it might be that the odors that we leave around the house when we leave *lessen* in a consistent amount each day. Over the hours that we are gone, our home begins to smell less of us. We could test this, I suggested, by bringing in a "fresh" smell of the owner. If the dog then assumes that the owner has just left, he will be surprised if the owner then returns.

That is just what happened. Working with a couple whose dog seemed to have magical powers of prediction as to when his owner returned, producers of a science program arranged to have a smelly t-shirt from one of them snuck into the house many hours after he left. Thus to the dog the house then smelled more strongly of the owner — as if he had left not long ago.

Sure enough, the dog was not waiting by the door when his owner came in, as he usually did. He was snoring on the couch: surely it would be hours before his owner returned, with that strong smell in the air . . .

Smell of Themselves

It seems reasonable to suggest that dogs know their own smell. And, moreover, that they like it: any owner who has watched her dog roll in the dirt after getting a hose-down can attest to their interest in not smelling like shampoo. On the other hand, does this indicate that they know it is *their* smell — or is it just that they don't like the green tea shampoo that you chose?

In other words, do they have a sense of themselves — of who they are? In animal cognition, this is the question of "self-recognition," which is considered a fairly complex cognitive ability — and which requires more than an observation of post-bath rolling to prove. The only test of self-recognition that has been reliably used with animals is the famous "mirror mark" test. In essence, if your face or body has been subtly marked or changed, will you move to examine the mark when you spy your reflection in a mirror? While we would expect that every normal adult would — and does, after eating spinach or poppy seed bagels — we are not born with self-recognition. But by the time a child is eighteen months old, he will reach up to remove a sticker an adult has surreptitiously placed on his head when he sees it in a mirror, thus passing the test. Chimpanzees pass (after being inked on their foreheads), an elephant named Happy passed (when an X of tape was placed above her eye), and captive dolphins pass (by doing bodily convolutions in order to examine the ink marks in reflective glass).

Dogs do not. Imagine showing your dog

the mirror when his face is covered with stickers. He will, no doubt, express indifference. What looks foolish to us is not of moment to him. But this is not sufficient evidence to say that dogs fail the test and thus have no sense of themselves. For one thing, dogs do not groom themselves (like primates) and show little concern for maintenance of appearance. So they are simply unlikely to want to correct an errant mark on their faces. Neither are they visually oriented as primates are. While the mirror test is appropriate for some species, this paradigm offers challenges for dogs, who show little interest in a mirror.

Some research hints that dogs might nonetheless be able to pass such a test, if a kind of *olfactory* mirror were designed: something that smelled *like* them, but was a small bit different. While walking his dog in the winter in the foothills of Colorado, researcher (and my colleague) Dr. Marc Bekoff wondered if every "yellow spot" in the snow was equally interesting to his dog Jethro. Bekoff began carefully noting where his dog peed and where his dog sniffed. He even ported some yellowed snow to new locations to see what happened. He found that Jethro avoided smelling his own urine but smelled others': a kind of recognition of himself, written in the snow. Jethro looked as if he were recognizing his smell, but I decided to put this hypothesis to a formal test. I wanted to see if any dog, smelling a reflection of himself, thinks, *That smells like me.*

To do this, my research lab and I set out to create a kind of olfactory mirror. Instead of a reflective surface, we used a canister exuding odor. When you look in the mirror, you see yourself. When you smell the canister, you smell . . . yourself. I used a dog's own scent as well as a revised "scent image": the scent, altered (or "marked"). We were asking whether dogs would tell the difference — and whether their marked self would be more interesting to smell.

This is how it happened that we started collecting pee.

One might not ordinarily think about handling, examining, or, perhaps especially, presenting urine to dogs. But pee is oddly central to our lives with dogs. Not only is it the great communicative medium between dogs; it is also a major part of the dog-human relationship. Perhaps you live with a dog in an apartment. Undoubtedly you must take your dog for a series of walks — or, at minimum, outings — for her to urinate outside. Should you be unable to get home during the day while working elsewhere, you might have a dog walker come and "out" your dog. If you live in a house, the dog needs to be let out in a timely fashion, or you need to design a way that he can let himself out.

My own social life as a young adult was to a great extent arranged around the fact that I needed to get home to take my dog, Pumpernickel, out to pee. Of course, I also wanted to give her the social companionship and exercise she needed and deserved as my loyal and bed-warming companion. But surely some of those outings were just about urine.

Then we leave it, of course. Dogs urinate, and we literally leave it.

Unless you begin to duck under their bellies as they squat — or reach around a raised leg — and stick a little cup in the way of the yellow stream. Then you gather it.

And that's just what we began to do. For if urine is so important for dogs, it is important for dog scientists.

Gathering urine is trivial work for a dog owner accustomed to handling a warm, soft pile of her pup's excrement with only the thinnest of plastic bags between skin and poo. Still, it takes some getting used to, for both persons and dogs. I established a methodology with my lab manager, Julie, who grew to have a proclivity for the task. Finnegan was our test subject. We went out for a walk together and both eyed him closely. He did not seem worried that Julie had latex gloves on and carried a sterile, orange-capped plastic cup. As Finn's owner, having accompanied him for thousands of walks, I could predict to a fare-thee-well when he was about to pee. Every owner, I suspect, has a strange body of knowledge about the set of behaviors that precede her dog's eliminations. So when I saw Finn take aim I raised my eyebrows at Julie, who ducked a cup midstream. Caught.

As we proceeded, there were considerations we had never imagined. First off, we needed to strategize the best way to scooch under a dog with a plastic cup without, well, the dog getting startled by a person reaching under him with a plastic cup. Some dogs balked at this — though they had given their urine away freely for years, they suddenly had proprietary inclinations. Or, more likely, that person's arm was just awkwardly close to their privates.

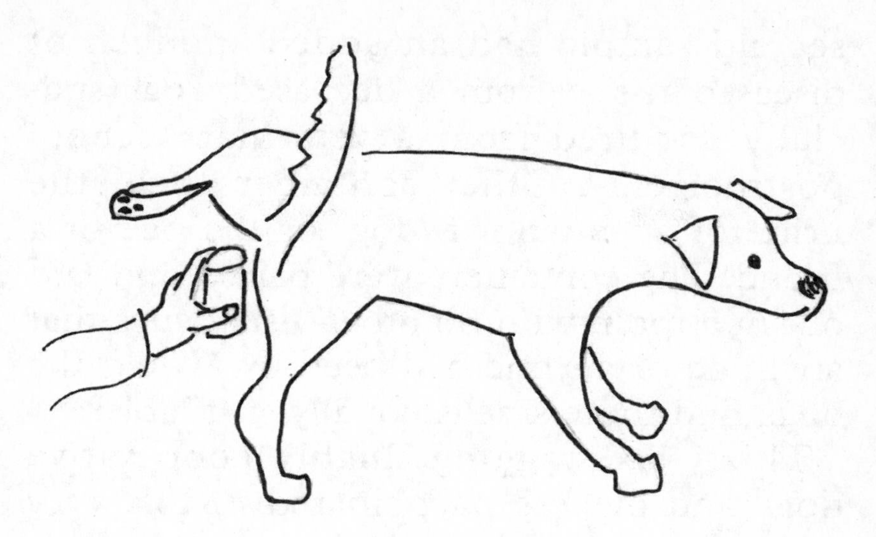

Then we had to determine just how much pee was enough to be smelled. We wildly overestimated on our first go: a small cotton ball dampened with pee. I called Finnegan over, showing him something in my gloved hand. He trotted right up to me. But on arrival, his reaction was quick and decisive. Have you seen the "disgust look" of a dog? A wonderfully clear judgment had been levied.

Eventually we got the proportions right (down to minuscule amounts of urine), and we also recruited owners to do the "materials" gathering for us. When they arrived at our experimental site, we made the first sample: a tiny bit of their dog's pee in a closed container with airholes on the top. A second sample had an added smidgen of diseased tissue from a deceased dog (specially acquired from a veterinary school postmortem). Other containers held the urine of an unknown dog, or the pee of a friend. The containers were paired and laid out in a room with no other visual cues that anything or anyone had been by. Would the dogs find these smells equally smellable?

Thirty-six wagging, highly cooperative dogs and owners participated in this very odd experiment for us. We have videos from various angles showing each nostril twitch, each eyebrow raise of surprise or alarm. We timed where they sniffed, how long they sniffed, and how many times they came back for another look. And the result: dogs passed the test. Our subjects spent way more time sniffing their olfactory image when it had been marked, as though looking at the mirror most when there was something funny between their teeth. They looked at (sniffed) themselves, to be sure, but not as much as when something was different about that smell. The smells of other dogs were also highly sniffed: as when another person looks over your shoulder in the mirror. You look back.

As in any experiment, some subjects offered additional behaviors, not falling under the strict rubric of the experimental design. Chimpanzees in the mirror mark task use the mirror to check out parts of their bodies they can't usually see: the insides of their mouths, their own rear ends, their nostrils. They make great faces at themselves. Our dogs did the equivalent. They scratched at and licked the containers. They turned back to their owners with looks of despair or excitement, sharing the news of this peculiar smell before them. Sometimes, I'll admit, they counter-marked the container. Any good dog cognition lab has lots of paper towels and disinfectant around. But they peed only on *other* dogs' containers, not their own. They saw themselves.

Smell of the New Day

Apart from the concrete — recognizing themselves or us — the dog's nose notices a world of abstraction. Why, to a dog, *the new day* has a smell.

Your morning dog walk may follow a route around the block, to the park, down the driveway and back. We take our dogs on these walks precisely because of their familiarity, their reliability. We humans are creatures of habit, and we assume that dogs are as habitual — and are experiencing the same habit, at that.

They almost certainly are not. Each departure from the house brings a new scene, one never visited. Each day, each hour, wears a new smellscape. For as people pass, shedding their scents and foods; as cars warm the street and line it with particulate dust; as the clouds coat the scene with rain pulled from alpine lakes; as breezes carry the scent from downtown uptown and the seeds from forest to plain; as the universe of animals, from bug to bird to dog, passes and leaves footsteps, excreta, and skin — the world outside your door is transformed. There is no such thing as "fresh air" to a dog. Air is *rich:* an olfactory tangle that the dog's nose will diligently unknot.

SMELL OF TIME

As each day wears a new smell, its hours mark changes in odors that your dog can notice. Dogs smell time. The past is underfoot; the odors of yesterday have come to rest on the ground. Carried by the morning's first breath or sloughed off the backs of nighttime animals, the message lies on the doorstep with the folded newspaper. The odor of the future is carried from around the corner, reaching the dog's nostrils before reaching our eyes. Smell rubber-bands time for dogs, pulling some of the past and future into the now.

Much of this information is on the breeze. Heads out the car window, ears sailing on the updraft of air, dogs *enter into* the breeze.* We are distracted by what we see of the wind: its effect on our skirts and hair, the flag's crisp responsiveness to the vagaries of the breeze. The dog experiences the wind from within, the messages it carries and the stories it brings from faraway climes. It brings word of a storm's arrival to the dog's nose before to ours. When low pressure arrives, and the air above the ground feels extra roomy, the earth loosens its grip on the odors it holds in reserve and begins to exude aroma into that space. Dogs, inadvertently trained by us to care about associations — to notice that "picking up the leash" precedes "going for a walk," that "scraping my plate" means "tidbits from dinner" — can easily learn that a bouquet from the soil means "a storm's afoot."

Smell of It All

Faced with bedbug or thunderstorm, the dog notices the details. You would not

*"The blind person enters into the windiness of the day at first hand," John Hull wrote about his experience of being blind.

recognize the smell of your house from your dog's point of nose. Yes, your home itself has a smell. With few exceptions, every item of furniture, every book and plate, every throw pillow thrown and table lamp tabled has a smell. Each exists in the visual world that we see, with a straight edge demarcating the space it occupies and the space it neighbors. But in the world of smell, the edge is fuzzy. It is clouded, and its cloud morphs from lamp to the shape of a bunny to train as it is touched or moved, as it is warmed by its light bulb quivering or cooled by its sleep.

It is not that dogs can smell each molecule of their environment. To appreciate their abilities one need not hyperbolize. But the "roseness" of a rose's odor profile may be less important, relevant, or perceptible to the dog than some of the other notes: citronellol and rose oxide compounds, say — a smell that sometimes is likened to metal or urine. Imagine being able to distinguish the parts from the whole, the past from the present, only with your nose.

How do they do it? The story of how those odors get into his mind, form the substance of his thoughts, and send his head upward in search of the telltale odor that his owner is approaching the door, begins with his konker, his snuffbox, his snout. Ladies and gentlemen, let me introduce you to the inside of the dog's glorious nose.

3: Sniffing the Wind

Upton is a scent-on-the-air dog. He stands and sniffs in the breeze for many minutes, his head proudly up, eyes with a faraway look, cheeks puffing slightly with every exhale. I almost feel I can see his moist nose, lightly spritzed with mucus, catching odorous words from distant callers.

To understand how the dog can do what he does, you've got to follow his nose. What a dog is experiencing is formed of what he is smelling: as we see images in our minds, he glimpses scents; as we speak in words, he communicates in fumes.

Have you toured the dog nose? Ridden on a corkscrew of air into the dark vault, bumped along its curves, caught a breeze up to the chamber where a molecule will settle into the wetlands and begin to tickle the nerves to the brain?

I have — at least, near enough for my liking. I've been down the nose in a simulation of an odor molecule's point of view as it is sniffed in. This unlikely video was created using airflow models generated by Dr. Brent Craven, who does work in computational fluid dynamics. Craven doesn't have a stake in dog olfaction, as such. His research into how fluids and air move is primarily basic science, geared to understanding animal systems. Its application (and grant money) is often to reverse-engineer particularly good noses, by using airflow models for the creation of artificial noses for military purposes.

The nasal airway Craven and his team modeled is actually that of an eastern gray squirrel — a perfectly good smeller, but "much simpler" to model, he said. Imaginatively superimposed with MRI images of the innards of the more complicated canid nasal cavity, though, the video provides a sense of what a bumpy, hurly-burly, complex journey it is from the point of view of the sniffed odorant.

You're riding on an odorant molecule. A miniature soap bubble comfortably bouncing and sailing on the briefest of wind currents, small and light and volatile. You're in the vicinity of a dog nose, then suddenly, abruptly closer. The nostrils gape wider and approach. The odor bubble is sucked inside. Its speed is immense. Terrifying roller coasters could be modeled on the beginning of a long-nosed animal's nasal cavity. For after a quick ascent, slowing as the path steepens, the odor's arrival at the peak brings a heart-stopping view: only air ahead. A rush downward, curving this way and that, slowing for a second, then resuming. New projections pop from the walls as the odor races toward them, forcing you to bounce to your side, hit your head on the ceiling, and drop with a stomach-lifting thump. There are convoluted curves, perilous edges, and, all along, a gravity-defying force pushing you ever deeper. A tenth of a second later (slowed down two hundred times in the simulation), you are summarily deposited in a marsh well stuffed with moist grasses standing soldierlike, awaiting your arrival.

And this is all before you get to the brain.

Nostrily

Finn snortles up close to me to wake me. I reluctantly peek out with one eye. Its field of vision is entirely taken up with Finn's right nostril, twisting this way and that, as we might do to make a silly expression. The result is that I twist my face into a smile and rise to greet him.

What has happened? At its essence, olfaction begins with the detection of as little as a few molecules of an odorant. The detector is the nose, the Hoover vacuuming in that trace amount. So let's begin there.

If you have a dog near (and I hope you always do) bend down and take a good look at the dog's nose. Get close, really close — one hopes, with a friendly dog who, at worst, will greet your examination with a peremptory lick of your own neb.

The hairy muzzle of the dog is what we usually look at — the snout, reaching back to the eyes and sloping on either side into jowls. Its length matters: while it isn't the part of the nose responsible for the detecting, per se, nor the sniffing, it is the well-designed hallway down which the sniff will tumble.

Or we look at the moist, hairless tip of the snout, the so-called planum nasale or rhinarium. This is where the action begins. Dog nose tips are mesmerizing: as different as thumbprints, and as full of individual details. What most wet nose tips share is the pebbled topography of their surface, polygonal cells made visible in their opal darkness. This surface is wet so that more odors can be picked up and absorbed into the nose. Moreover, they have temperature gauges that can direct them toward a cooling breeze that might carry odors.

And their nostrilness! The nose tip carries two gaping nostrils — or *nares* — which will lead to the *real* nose an inch or six north, where special tissue lines the nasal canals. For those who find it hard to love a nostril, I say: You haven't spent long enough looking at your dog's. While to the uninformed they are just "holes in the front of the nose," to those who study how dogs sniff, they are "articulating aerodynamic inlets," their bulbous shape facilitating high airflow. Ringing the nostrils are the highly developed muscles in the alar folds. It is this anatomy that allows the nostrils to be an active part of the sniff, and that gives nostrils such a terrific range in shape from dog to dog. Many look like commas curled tight, or rounded into disks. Others are mere blunted openings, as if drawn hastily with a large-tipped pen. In some countries a noseprint is used as an identification: the inexpensive microchip. Ink the thin outer layers of the nose tip and press it against paper. A dog's noseprint is the newborn's footprint all parents will stare at for some clue as to their progeny's future.

Dogs can use their nostrils separately and differentially: when sniffing something new and “nonaversive” — neutral or likeable odors — they begin with the right nostril, then switch to the left. By videotaping down the length of a cotton swab dabbed with an odorant while dogs sniffed, researchers found that lemon, food, and the secretions of a female dog elicited this right-then-left sniffing. When faced with the smells of adrenaline or the sweat of a kennel veterinarian (who swabbed his armpit for the sake of science), though, dogs sniffed *only* with the right nostril. It is thought that this nostril preference is due to the side of the brain involved. The right nostril connects to the right hemisphere of the brain (ipsilaterally, in contrast to the other senses, which cross sides) — which is more often associated with fear or aggressive behavior than the left, which analyzes familiar stimuli. If a dog is sniffing you right nostril only, he may be feeling suspicious.

Back to those muscles around the nares. Outside in a mild breeze, I watch as my dog Upton turns his nose — just his nose tip — to the right and left, bending in a way you don’t think a nose should or will bend. His nose gymnastics get the nostrils facing the breeze, picking up turbulent air, and snagging an odor. Then they open the aperture of the nostrils wide to maximize the amount of air that is pulled inside. An odor looks inside the business end of the telescope that is the nostril and closes its eyes for the ride.

SNIFFER

When I return to the house after an absence, I bend to Finnegan to let him sniff me, and to let me hear him sniff me. His sniff-snuffle is almost communicative — nosed phonemes in strings of merry sentences.

What we know about what happens when the odor faces its future up the dog nose comes largely from the work of mechanical engineer Dr. Gary Settles, one of Craven's former professors. Settles, now emeritus at Penn State, has brought fluid dynamics — a field associated more closely with designing smooth-flying airplanes — to the study of noses. To my great delight, what Settles studies is the *sniff.* He and his team ask dogs to wear a specially designed muzzle, give them good things to smell, and measure the fluid dynamics of their sniffing.

Settles speaks of the dog's nose as a classic "variable-geometry aerodynamic sampler." The "sampler" (the nose) approaches a vapor cloud (air with smelly stuff in it) and transfers that cloud into its internal sensor chamber (the back of the nose, where the neurochemical magic happens).

The transfer happens through sniffing. As an invisible means of pulling in what are invisible odors, the sniff has long been downplayed. Further, while vision happens through light "hitting" the eye, in olfaction, odors do not just hit the nose; they sail right in. For that reason, smell feels intrusive. It is common for people to suggest that this is so because if one is alive and breathing, one can't help but also be smelling: a gulp of air enters the nose with each breath, carrying, presumably, various odors. While it's true that a nose-sized swig of air is bound to have smellable stuff in it, it's not true that smelling happens by being alive and having a nose. It turns out that to smell, one needs to sniff. Not just breathe, not just sit with nostrils open. In the mid-nineteenth century, the start of the hey-day of scientific discovery based on self-experimentation,* the physician Ernst Heinrich Weber, a

*Practitioners included Freud, who famously experimented widely with cocaine; plenty of others who ate or drank questionable (including radioactive) materials; and researchers who self-catheterized or gave themselves viruses or unfinished vaccines.

founder of modern experimental psychology, first demonstrated the importance of the sniff. Weber lay down, had a solution of water and cologne poured into his nostrils, and waited, motionless. This is just the kind of experiment for which it is hard to recruit a lot of volunteers. What Weber found was that the normally ripe cologne was undetectable: he could not smell it when it was passively introduced to his nose.

Weber subjected himself to a noseful of cologne in the interest of discovering whether smells need to be delivered in vaporous form or can be detected in liquid. His was not the final word. A hundred years later, researchers renewed experimentation on the means of smelling: some repeated Weber's supine nose-filling experiment; others presented strong smells before a person blowing through his lips like a trumpeter; and still others tried injecting odors intravenously to people during sleep apnea, between breaths. These studies had one common design element: in no case could the subjects sniff. The results? None of the bemused subjects perceived any odor. During typical breathing, only a very small amount of the air you inhale makes it up to the part of the nose holding the cells that will receive and manage the smell. One simply must energetically sniff air in.

So sniff, already! Sniffing is the inhalation part of a breath (you sniff in, but not out — that part's more rightly a snort), often audible, requiring a modicum of effort. The variety of sniffs in the animal kingdom is certainly underplayed. Elephants sniff strongly, with trunk hovered over a scent or raised in the air for a "periscope" sniff. Gerbils sniff rapidly, noses twitching; tortoises, by contrast, have a slow-motion exploratory sniff in which they extend their necks, orient their heads down, and open their nostrils. A small puff of dust may appear on exhale. New Guinea singing dogs even "huff-sniff" when pursuing prey such as voles in a hole or under vegetation, blowing vigorously out the nose before sniffing vigorously in.

When canids first appeared, some tens of millions of years ago, their sniffing was enabled by the bellows that were the lungs, probably through a fairly straightforward nose. For fish, frogs, and reptiles, the snout is just a cavity allowing water and air to wash directly over the olfactory cells. This is still more complicated than the current invertebrates, many of which smell a lot, but not by way of sniffing smells *in.* Invertebrates have olfactory detectors outside their bodies, such as on antennae, so they must actually run into the odor source. Some — like lobster — send their sensory organ forward to sniff; others sniff simply by flying or wiggling their bodies and letting the air or water hit them.

Not for dogs, that crude method. Dogs not only have multiple sniffs, at different rates and for different purposes, but they have an ingenious method of *exhaling* that aids their smelling.

Settles discovered how dogs sniff by watching what he quaintly calls "canine nares airflows." He recruited a handful of pets as well as detection dogs to come into the lab to be observed and recorded. There were golden retrievers, Airedales, Labs, German shepherds: all manner of noses, trained and untrained. The researchers put a variety of objects — highly desirable food, some inedible and novel scents, even a tiny bit of TNT and the sweet smell of marijuana — either right in front of the dogs or a distance away. Without being asked twice, the willing subjects sniffed. Detection dogs and pets sniffed alike, whether they were presented with pot or pet food. Both groups had two kinds of sniffs. When the smelly stuff is far away, unreachable, dogs do a "long sniff" — a highly directional, air-pulling sniff that lasts two seconds. In the long sniff, nostrils dilate and the alar fold opens; the mouth may be slightly open. Imagine the regal-looking large dog on a hill, his chest forward, his nose in the breeze. That's the long sniffer. This sniff is often the culmination of a bout of sniffing, many weaker sniffs followed by a whopping one. In fact, one English pointer skilled at hunting birds was found to be able to maintain an extra long sniff. "Sir Satan" — the only of the tested dogs who was willing to suffer a sensor fastened into his nostril — maintained a continuous inhalation through his nose for forty seconds while running into the wind after a scent.

On the other hand, when the smell is nearby and on the ground, dogs sniff in bursts. First, they scan the surface. Have you ever noticed your dog nosing for a toy in a grassy area and seemingly *not finding it,* though she's right on top of it? Her nose isn't deficient: she is simply doing a survey of the area, as we scan a scene with our eyes. Typically a dog will walk her nose right up to the toy, actually pause over it, snort a little exhale, then continue on. To the vision-obsessed bipedal creatures nearby, it seems she's missed it — rather stupidly — but she has not. She returns to the toy. She is simply assessing the concentration of all the smells in the vicinity to find the strongest source — something like a visual inspection of all the brunch buffet selections before you zoom in on the Belgian waffle (that you probably knew you wanted all along). Along the way, the dog does rapid "short sniffs," from five to twelve every second, enough to make me hyperventilate just imagining it. This sniffing occurs at about the same rate as a dog's panting — 5.3 tongue flaps a second, on average — which is energy-efficient, and so might be their sniffing.

A hemisphere of air from four inches away is pulled up and in — this is called the nose "reach." Given the chance, as every horrified dog owner knows, dogs will work to decrease that reach to a centimeter or full-on contact with the odor. Owners sometimes ask me, *If their sense of smell is so good, can't they smell it from here* — i.e., at a polite, safe distance? But we mistake their noses for ours. They are not trying to detect it at all; they are trying to discern its contours, to take in all of its features, to make a measure of the smell.

Indeed, it is at a centimeter out that the smelly air is drawn in fastest. At this distance, the dog can get different odor samples from each nostril, which wind up giving the dog a bilateral vision that is a kind of "stereo" olfaction. Just as the images from our two eyes are constructed into a three-dimensional image of the world, the differences in the strength of the smell image in each nostril help the dog locate the source of the smell in space, whether it's to his left or right, fore or aft.

Given this, asking why a dog is *sticking his nose right in that dog's rump* is the equivalent of asking why you like to experience van Gogh's *The Starry Night* from close enough to see each brushstroke — rather than peeking in the doorway from the next room.*

Canine Air Jets

But the most profound difference between the dog sniff and our own sniffing comes when dogs exhale. Our exhalations go straight out the entrance, out through the in door, pushily shoving any new air out of the way and preventing it from getting in. This can be a terrific relief when you want to purge a horrible odor from your nose, but it also sends lovely odors out almost as soon as they have arrived. When a dog exhales, he creates what Settles calls, charmingly,

*Yes, I am equating the dog rump with *The Starry Night*. Dogs have perfectly artistic rumps.

"expired turbulent canine nostril air jets." Through high-speed videography of nostril and air motion, Settles found that dogs create tiny wind currents by exhaling not straight out, but out the *side slits* of their noses. This strategy minimizes the odor displaced — what Settles calls the "sample blow-off" — by the puff of air. The wings of the nostrils flare, a nose-plane ready for liftoff, and expired air leaves through a sneaky side exit. Not only does it not push the odor out of the way: the exhale creates a puff that lifts *more* smelly particles off the surface and a suction that hurries the next noseful of smell inside the snout. The exhaled nostril jets are little rotating funnel clouds that pull Dorothy, her house, and her little dog, too, right up into the nose.

Remember your dog's short, thoughtful

pause over the toy she is seeking? This is a pause of moment. She is sending her "expired air jets" right onto the source. More clouds of odorous particles come up from the toy and the ground. These jets are essentially increasing the reach of the nose, blowing and vacuuming in sync.

One scientist I spoke with likened this kind of sniffing behavior to the circular breathing a player of a wood or brass instrument might learn to do. It is the sniff without punctuation, allowing dogs to get a continuous read of the world — just in the way that we see the world without pauses while we blink.

Settles saw this jet-enabled sniffing by using specialized video called schlieren photography, which uses mirrors and slow-motion cameras to capture a picture of airflow. The photos make warmed air visible as blurred clouds emanating from the nose and mouth. In slow-motion schlieren videos, dogs' snouts turn nearly prehensile, reaching and retreating to move the air; the muzzle seems to undulate, like a jellyfish pressing himself through the depths. But you can see some of this with the naked eye, too: simply watch your dog sniffing over a dusty plot of earth. After a particularly enthusiastic bout of sniffing some invisible thing, you can easily spot the puff of dirt and dust and odor pushed into the air and into the dog's face with his own snorty exhale.

Could a sniff really be responsible for dogs' keen sense of smell? Well, it is one of a number of key components. We know this because of evidence from when dogs lose their nose: when they are panting. Hot dogs can't smell much. Dogs do not have sweat glands allowing them to release heat through pores across their skin. Dogs have only their waggy, pulsing tongues. They must pant — and panting, schlieren photos show, forces out such a wash of air that no smelly air can get to the nose. The panting dog must close his mouth to sniff well.

My Finnegan's sniffing is audible, his own particular combination of huffing and grunting and spluttering. Outside, he is a nose-on-the-ground dog, trailing invisible stories in the grass. Give him a tree trunk that is popular with the local dogs, and he will serenade it with his rapid huff-splutter and his turbulent-jet exhales. At the other end of the leash, he will bring me to a stop with his nose set into an odor, his whole body cemented to the ground to allow the nose free smelling. Our Upton, the second dog, *learned* sniffing from Finn — learned that it was fine in this house to stand fast and observe the scents before moving on.

SNOUT

If one didn't know which sense was most dominant in humans, it would take only a few minutes of observation to discover. Everything we want to perceive or examine, we parade before our eyes. Something to the side? We turn our heads (moving our ears away) so our eyes are nakedly upon it. Hear something overhead or underfoot? We don't try to listen up or smell down: we turn our eyes. Our faces, while giving equal billing to the nose and mouth, have an array of protective devices around the eyes — eyebrows, top and lower eyelid lashes — and many of us use our noses simply as a perch for large spectacles to help us see better. *Do you see?* we ask, as a stand-in for understanding. Not *Do you smell? Do you taste?* Seeing *is* understanding, for us. When we meet each other, we greet each other with our eyes: to not look is considered rude, even abnormal. When walking, we start to turn our heads — and eyes — toward a corner seconds before we direct our legs around it.

So, too, with the dog's nose. There is a fair bit about its simple position on the dog's body that enables finely tuned sniffing. The snout is plainly protuberant. Not by coincidence, it is at the end of a head which, via a highly flexible neck, can reach the ground — where most odors lie. Dogs don't spend a lot of time worrying about sniffing treetops: they sniff objects emanating from and landed upon the earth. As well as parts of other dogs' anatomy that are, well, about nose-height.

Dog snouts are long for a good reason: evolution commits large amounts of anatomical real estate only to very useful causes. The cavelike, moist interior vaults under skin and fur are filled with air filters, humidifiers, and warming devices. When you're sniffing dog rumps and decaying squirrel carcasses for a goodly portion of your life, you'd better have an excellent air filter. Inhaled air is cleaned and conditioned before it gets to the back of the nose. It must be dressed and ready for courtly presentation, as in the back it will meet kingly neurons that travel directly into the brain.

"So, you've been caught by a nostril," Craven says. He's walking me, riding that intrepid odor, through the interior snout. "The flow is pretty high in the vestibule," the antechamber when the odor gets in the nostril. The dog is pulling in with his lungs through the nasal pharynx; the air is turbulent and chaotic — and then there is a fork in the nose. Inhaled air can go one of two ways: the breathing route or the sniffing route. If you go the respiratory route, you are heated up and humidified and proceed to the lungs. But if you're sniffed to be smelled by a dog, you take a whole 'nother route, a high-speed ride toward the olfactory region. A current of air zips along a labyrinthine, tortuous path past a series of thin, curvy bones called turbinals; in cross section they resemble a big brain, convoluted and branched, folded into a small space. The turbinals, too, are part of the cleaning system — and, farther back, some are also covered with tissue useful for smelling. The turbinals create the roller-coastery pathways along which the odor rides on its sniffed journey.

Since air flows only one way across the long olfactory road up the nose (odors then are exhaled via the breathing route or are broken down by enzymes), dogs may be able to do something extra fancy with their noses: sort smells into groups as they fly in. Some odors are absorbed more readily than others, which means they will be grabbed by the sensory cells earlier in their route through the nose. For instance, researchers found that one component of the explosive TNT — DNT — is snagged more readily than other odors; this may partially account for sniffer dogs' seeming ease at identifying it. It is more soluble, so dissolves earlier in the nose, than a molecule like amyl acetate (smells of bananas), which is itself more soluble than limonene (smells of lemon), which, like many other kinds of odors, will make it back to the olfactory recess before being absorbed into the mucus and finding a receptor to bind to. While we rarely ask dogs to find bananas for us, we could.* Given the layout of different sensory receptors throughout the nose, a dog can start to identify and discriminate smells in his nose before the brain even gets involved.

Toward the back of the channel the odor suddenly slows. The turbinal bones here are lined with olfactory epithelium — brown tissue that holds the sensory cells, which will snag odors out of the air and is where the magic conversion from "odor" to "odor you smell" begins to happen. The cells are

*And, in fact, there are now many detection dogs trained to find bananas — as well as any other agricultural contraband brought across the border illegally.

coated in mucus.

Then, “At some point, you will be deposited or absorbed by the mucus that’s lining these airways, and you’ll slowly diffuse,” Craven warns. It’s not as grim as it sounds. A thin lining, about ten microns thick, forms the transition from the outside air to the internal neurons. The odor you are riding will migrate through the mucus in a tenth of a second, about the time it takes for a sniff to travel from the nostril to the back of the nose. But for now you can relax.

Recess for Odors

You’ve arrived at the cul-de-sac of the snout, the olfactory recess. This is the part farthest back in the nose, at a point a half inch into the skull, just between the eyes. In the olfactory recess odors can linger, trying to find a sensory cell to nuzzle up with for many rounds of inhaling and exhaling. Dogs get a chance to really ruminate on the smells they’ve sniffed in before the air is scooted out.

The recess — as well as some of the bones along the way — is lined with the aforementioned epithelial tissue. When newspaper articles say that a dog’s sense of smell is ten thousand times better than a human’s, or one million times better, or whatever-exponent-of-ten times better, one of the most cited bits of anatomical proof is the amount of olfactory epithelium — the expanse of nose with cells specialized for smelling smells. Though the comparative numbers are suspect (and quite variable for different odors), dogs are massively better endowed than humans in olfactory sensory cells. If his olfactory epithelium were spread out along the outer surface of the dog's body, it would completely cover it. In humans, ours would about cover a mole on our left shoulder.

The epithelium here is covered with a dense mat of cilia, little hairlike branches that extend from the sensory neurons. A few dozen cilia sprout from each nerve, and each is coated with dozens of proteins called olfactory receptor cells. Receptors do what they sound like: they receive odors. To do so, they stick, untroubled, right out there in the mucosal environment of the nose, perfectly designed to snag odorant molecules coming by on a sniff.

Dogs cram more smellability into their snouts at every level: they have many cilia on each neuron and more receptors on each cilia than humans do. Indeed, every dog has *hundreds of millions* more cells devoted to detecting smelly stuff than humans do.

Dogs have from two hundred million to one billion receptor cells, depending on the breed, compared to the six million in our noses. In the dog's case, more nose mass also enables more *kinds* of receptors — over 800 — that themselves can encode more information about the odors.

This number — eight hundred and change — gives researchers pause. The eye, conveyer of the brilliance of a sunset dancing off the clouds after a thunderstorm, uses just *three* receptors to draw this colorful scene in our heads. With eight hundred more receptors, the possibilities for the odor landscape are staggering. The number of odors the dog could detect could theoretically be "billions," Dr. Stuart Firestein, a neuroscientist at Columbia University who studies olfaction, has written. But "in fact, the question is probably not relevant, just as it makes little sense to ask how many colours or hues we can see."

Even when odors land in their receptors, they are still undercover. The nose does not know what they are. There is no "cheese" receptor, activated by the Stiltons and cheddars my dogs nose out on our kitchen counter. Despite the dog's alacrity in locating the corpse of an unlucky squirrel in the park, there is no "dead squirrel" receptor.

Simply, each odor activates many receptors; there is not one receptor for each odor.

Although the means of reception has not yet been conclusively determined, the most popular theory of how it happens uses a lock and key metaphor. In this model, receptors are the locks, of various shapes and lengths, and the various molecules that make up odors are the keys. A related theory suggests that the reception of odors is less specific than a key in a lock; instead, it's more like a key in a pocket, in which many differently shaped keys can bind to a receptor and prompt it to fire. Apt for dogs, smeller of pockets.

When biologists Drs. Linda Buck and Richard Axel won the Nobel Prize for their work in olfaction, it was for discovering the genes that code for these receptors. Amazingly, olfactory genes are wildly overrepresented in the mammalian genome. Dogs have around eleven hundred olfactory receptor genes, some eight hundred of which are operative.* Keep in mind that

*The rest are "pseudogenes": genes that have a mutation such that they no longer result in the development of a receptor — that is, they no longer make the thing that they are supposed to. For dogs, 20 to 25 percent of their olfactory recep-

your dog's genome — the blueprint for making his entire body, from his charismatic curled tail to dark expressive eyes — is just over 19,000 genes. Nearly 5 percent of the genome is committed to making smell receptors alone — plenty of kinds of locks with which to smell the world's keys (or pockets).

Breeds vary in their smelling abilities, and it may be because they have varying numbers of operative olfactory genes: boxers (short-nosed, compressing that turbinate-filled snout) have slightly fewer functional genes than poodles (long-nosed and respectable smellers). Though research has just begun on the topic, there is some evidence that specific genes might even be linked to detecting specific odors. In one study, a tiny substitution on a particular gene was seen in dogs who did less well on explosives-detection work.

If genetic differences lead to differences in detection ability, one may well ask, does this mean that certain breeds are genetically superior sniffers? Insofar as some undetermined subset of a dog's genes may lead him to notice a smell that another dog does not,

---

tor genes are pseudogenes. For humans, it's more than 50 percent.

yes. Whether the genetic difference in smell abilities is part of the genetic difference in breeds is another question — an unanswered one, for now.

The Smell Hits the Fan

All that work to get to the receptor cell — but the odor is not done yet. The smell of dead squirrel — or any other substance — is discovered only when components of its odor, having snuggled into receptor sites, cause the neurons to fire — changing voltage — and sending an action potential down its length, leaving the nose and entering the nose brain: the olfactory bulb. Tens of millions of neurons converge into a few thousand bundles and sneak through small openings in the bone into the brain. Long ago, it was thought that brains did the smelling, and the nose was just the conduit. Even in the twentieth century, with the rise of brain/computer analogies, it was not uncommon to hear the nose be called simply the fan of the massive supercomputer that is the brain. Now we know it's a bit more electrical. Santiago Ramón y Cajal, an early and influential anatomist, mapped the route from the nose to the brain at the end of the nineteenth century, observing that nerves (carrying word of odors smelled) enter the brain, not the odors themselves.

The olfactory bulb sits right behind the back of the nose, jammed in under the frontal lobe. The nose is the quickest route to the brain: one neuron goes from the warmed, dinner-smelling, light-spritzing-of-dog-hair-wafting air in your living room to, on the other side, the highly conditioned environment of the brain. Before the olfactory bulb, all a neuron "knows" is that it's firing; in the bulb, axons from thousands of neurons from the same kind of receptors all converge on a single target site and flood it with activity. What this seems to enable is the cobbling together of a smell sensation. Just as an odor molecule is grabbed and deconstructed by the receptors, causing many to fire, it is reconstructed in the topographical layers of the olfactory bulbs. Little traces of evidence from many cells are translated into the fetid, cadaveric smell sensation that is your dog's discovery.

One might expect the dog's olfactory bulb to be humongous. It's not. But it is 2 percent of his entire brain, two pennies to the silver dollar of the brain. (In humans it is vanishingly small: less than a thirtieth of a penny.) And this matters: the olfactory bulb translates those receptor cells firing into something like an experience of an odor.

From the olfactory bulb the odor information is zipped through the dog's brain on a massive and near-instantaneous search for recognition, how to feel about the smell, what memory it might awaken, and what behavior it might prompt. The bulb connects straight to the olfactory cortex for some of these decisions, as well as directly with the subcortical limbic system, which adds the emotional tone — fear, excitement — to the odor.

Research on the brain's response to odors has always had one common element: present the subject with something truly stinky. Measure the result. An early researcher studying the brain's response was physiologist Edgar Adrian, who exposed the humble hedgehog to various stinks, including the odor of decayed worms. "Water in which an earthworm had been allowed to rot" prompted a big reaction in the hedgehog.

More contemporary researchers have presented rabbits with store-bought sharp cheddar, lured tsetse flies with a combination of oxen and buffalo urine, exposed terrified rats to anal gland secretions from weasels and red foxes (rat predators), and, most recently, stuck the smell of their owners' armpits under the noses of dogs.

To back up. Neuroscientists can be fond of trying to answer the big questions about the mind with machines. MRI machines, when you are in one or looking at images generated from one, can seem superficially like mind reading. In "functional" MRI, a subject lies on a platform surrounded by walls housing a very powerful magnet. Through disturbance of the magnetic field, images of the blood flow in the brain (indicating neural activity) can be captured. Lie down, think of your grandmother, and the areas of the brain where your memory of her crooked smile and glasses, the smell of talcum powder, and the melancholy memory of the toys she made for you as a child reside will light up on the computer screen.

fMRI will never answer the questions of *what it is like* to have those memories or to revisit her smell. Images of brain location do not explain how the odor of talcum prompts me to remember my grandma sitting in her special chair in her dark, stuffed living room. Instead, the device lets us view the activity from afar: watching the Perseid meteor shower without answering the questions of the universe. And so a few research programs have begun looking at dog brains in MRI machines. This is itself a feat of training and patience — though quite doable for both species — as the dogs need to be very much awake in the machine, and yet stay perfectly still. One early study looked at what areas of the dogs' brains were triggered by their owners' smells.

By "owner's smell" I mean "smell of the person's armpit," caught on a gauze pad and waved in front of the dog prostrate in the MRI magnet. They found an area called the caudate nucleus dancing with excitement at the armpit gauze. Researchers had looked there because it's easy to visualize in the MRI, and it is associated with reward.

I suggest that another approach to discovering where the olfactory information goes in the brain is to watch your own dog. If you live with a dog, you know what happens after dead squirrel is identified by the brain. No dog waits and considers the wisdom of his next move before doing it.

The dog rolls in it.

This jibes with a theory Stuart Firestein advocates. "My craziest idea, you want to hear it?" I do want to hear it. "I think the actual percept of an olfactory object" — the sense of *experiencing* a smell — "may not occur until you hit something very close to or actually in the motor cortex," Firestein tells me. "Because most of what we do in olfaction is very tied to decision-making: do you put this in your mouth, do you flee from it, do you fuck it, do you whatever-you-want-to-do-with it?"

Indeed, some signals from the olfactory bulb do go straight to the motor cortex, a part of the brain that controls movement. It directs the muscles in the dog's legs that cause them to buckle gracefully, the precise angle of the head to squish perfectly into the smelly stuff, the intensity of supine squirm necessary to distribute the scent down his back. Note that when the dog gets up from throwing his body on and thrashing about in that squirrel, the first thing he wants to do again is get another whiff.

VNO

He licks. He licks my knee, my face, my ear. He licks the air, tongue reaching for me, as he approaches. It feels like affection, and I smile. But he also licks the corner of the building, the very smelly-looking patch of grass, and — sigh — the cat's bottom.

So much nose! But in dogs, and in many mammals, the olfactory system is bigger than just the nose. Dogs have a kind of "second nose," right under the bone separating the nostrils and above the roof of the mouth. Two scrolls of cartilage house what's called the vomeronasal organ — snappily abbreviated to VNO by researchers wanting to save themselves four syllables — which is as much a part of smelling as the nose is. Despite its evocative name, *vomer* describes the shape of the organ: ploughshare-like, the part of the plow that does the cutting. Secreted under the nose, the organ cannot be reached by simply sniffing; instead, odorants must dissolve into the tissue and be sucked inside. The pumping mechanism is prompted either by touching the molecule directly or through making a horrifyingly silly face called flehmen. Should you see a horse curl back his upper lip, seeming to scowl and shuddering a bit, you are witnessing classic flehmen: it pulls the odors back to the nasal tissue to be absorbed. To flehmen, the pig opens his mouth wide; the cat holds her mouth a small amount open, giving her a disconcertingly muddled appearance. A snake's forked tongue flicks to pick up odors and send them to each side of its VNO.

Most dogs do not do a classic flehmen lip-curl, but they have their own methods. At times, after sniffing, a dog may wrinkle his nose in an apparent grimace and chatter his teeth — that's a canine version of flehmen. Even better, they lick. The dog's extravagant, long tongue is good for reaching the bits at the bottom of the peanut-butter jar and cleaning your legs of post-exercise sweat, yes, but it is also a perfect mechanism for bringing odors to the VNO for investigation. Lick ground, lick nose, *smell.*

What the VNO enables dogs to detect is a kind of molecule that the ordinary olfactory route often cannot, such as pheromones. A pheromone was originally defined as a signal sent between two members of the same species — which causes the receiver of that signal to behave or develop in a very specific way. Androstenone, produced by a boar, causes a sow to assume, rather robotically, a mating posture; bombykol, given off by the female silk moth, wafts onto the antennae of the male silk moth and causes him to seek her out. Pheromones are used by an incredible diversity of organisms, from lobsters and rabbits to ants and bacteria.

What makes these pheromones detectable by the VNO is that they are typically water-soluble chemicals, nonvolatile, low-molecular weight molecules. So are lots of other molecules, such as hormones and "signature mixes" that may hold information about an animal's identity or the particular family or pack it belongs to. The VNO receptors are tuned to be highly specific and sensitive, unlike the more broadly receptive olfactory receptors in the nose. The rump-to-face-to-rump sniffing dance between your dog and another dog is a chemical communication of their sex, readiness to mate, their health — and also who each of them is. Each animal's urine and saliva may hold the same.

SNEEZER

Apart from the snout proper, the featured player on the face, the dog's sense of smell commandeers other dog features and behaviors for its olfactory ambitions. Not for nothing are bloodhounds so well ear-endowed. James Thurber's famous sketches of the breed have as much ear as head. And ears help make members of the breed such outstanding smellers. With his nose pointed down, the bloodhound's long ears — over thirteen inches, in extraordinary cases — sweep up odors from the ground for better smelling. They are dual fans attached to the face to direct possible smells to the smeller. Even the drool that hangs from a bloodhound's mouth might help to bring odors upward to be absorbed by the vomeronasal organ for consideration.

I discovered two other ordinary behaviors involved in smelling through simply watching dogs. Many of my research programs with the Dog Cognition Lab are rooted in observations of dog behavior in natural environments — for instance, in public parks, around people and other dogs. I have spent untold hours videotaping dogs playing together, then coding their behavior by reviewing the tapes one-thirtieth of a second at a time. I code the video in this super-slow-motion way because it enables me to see things we miss in real time. But there is another benefit: watching video at that pace defamiliarizes the dog. While it is terrific fun to watch dogs playing (believe me, it was hard to convince a dissertation committee that this might be actual work), our natural human tendency is to see the play but not see what is really going on. To see, but not see.

When we see dogs, we are so immediately and confidently sure of what they are doing — *Oh, they're friends; look, he wants to join the play; she is shy* — that we preempt actual examination of what they are doing.

It was while out watching dogs in this way that I made a surprising observation: dogs wag their tails a lot while meeting someone new or greeting a dog they know.

Wait, you say: *that is not new!* Of course. We *know* what tail-wagging is for. A loose, jangly, high tail-wag is a sign of friendliness. A fast, low tail-wag indicates anxiety. This is true — but simply knowing this makes it harder to see what else tail-wagging is doing. It is spreading scent. Whether intentionally or not, when a dog wags her tail, all the very fascinating odors (to a dog) from her anal sacs spread in a bloom around her body. She is not only telling other dogs how she's feeling; she's scenting who she is.

There is precedent for the use of the tail-wag as scent-spreader in other animals. The hippo madly fans his tiny tail while urinating and defecating, better to atomize that scent. Some rodents and squirrels wave their tails when a possible mate is nearby. Woodland caribou, whose tail is covered with scent glands producing a musky odor, use it as a "scent brush" to disseminate an alarm odor. So though a dog wagging his tail looks simply like a dog wagging his tail happily, he is also saying *Come smell the wafty odor that is me.*

Similarly, walking with my dog Finnegan one day, I noticed another usual but unusual behavior. As we passed a tall, prancing black poodle being walked on a tight leash across the street, our dogs looked at (smelled at) each other. Suddenly, the poodle did a full body shake — while still walking. I could very nearly *see* the plumes of scent from her curly hair hurtling toward Finn's nose in a fume cloud, and he was fixated. *Could a shake be deliberate?* I wondered. It might be the canine equivalent of flirtingly running my fingers through my hair as I talk with a suitor, perfuming the air around me with the scent of some slightly overpriced shampoo I use.

Trying to see through the familiar dog behavior also allowed me to learn about the sneeze. Sneezes are sneezes, to be sure: reflexive ways to clear the nose of something ticklish or alien. Dogs sneeze only through their noses, as opposed to the human mouth-nose sneeze. The nose-only route marks an interesting secondary use of sneezes: to clear the nose of an undesired smell. In this case, I believe dogs use sneezes intentionally to end a session of smelling one odor and to ready themselves for a different one. Watch your dog sniff a strong, recently laid smell at a corner, then see if she doesn't sneeze before moving on down the street. Relatedly, trainers of tracking dogs have noted that some dogs clear their noses by lifting their heads up from the ground, taking in a bit of less scented air. It is not a pause or a miscue; it is an essential part of his tracking.

In the end, it is not the size of the olfactory bulb itself that makes the dog's nose so keen. It is not the numbers of receptors per se, or the way that the dog sniffs. It is not just the length of his nose. It is *all* of it. It is sniffing in the way he does, through the nose he has, into the many receptors he houses, to the brain that has evolved. The result is astonishing.

You have surely seen the sleeping dog who, eyelids fluttering, toes dancing, barking muffled through a closed mouth, seems to be dreaming. Dogs do have REM sleep — the time when we most often dream — so it is likely that they are also dreaming: chasing, with those dancing toes; announcing, with that muffled bark. Next time you see your dreaming pup, look for his active nostrils. For a creature of the nose is almost certainly smelling in sleep — following the odor of his friend down the street or of dinner newly plopped in a bowl, or investigating a curious smell arriving under the door.

Go and look at your dog's nose again. I love to look at my dogs in their eyes — a shared glance is full of understanding, an agreement that we belong to each other.

That dogs look at us is what bootstrapped an ancient proto-dog, lurking around the edges of early villages, into the dog who sits on your lap and wags at the sight of you.

But now I look at my dog's nose, too — the whole length of it, but especially its wet tip — and my heart also skips. What moves me is the depth of what that nose knows about the world.

Talk to human psychophysicists and neuroscientists, and they'll say, *You know, it's a myth that dogs smell so much better than humans do.* And you can look at them and ask them what they smelled today. The answers vary from "nothing" to the poetic: "grassy meadows, prairie grass," smell scientist and author Dr. Avery Gilbert reports to me, of his new home in Colorado. Pretty good. Now consider asking a dog. If a dog could talk, his answer would be an epic poem recited over several hours. In fact, humans have fine noses. But most of us simply don't bother to smell. I mulled this over as I sniffed Finnegan's fur one night (smell: fresh running river, water burbling over river rocks). Hmm. Maybe I'd bother to: to try to make my nose a worthy companion to my dogs'.

4:
Walking While Smelling

Most dogs are not employed to find bedbugs or spot cancers. Most live in owner-controlled smell environments, asked to do little smelling beyond locating and cleaning up any food the resident child has dropped from his high chair. Moreover, in dogs' peculiar indoor world, the smells of the day that accumulate on hardwood and carpet are periodically vacuumed up (which period depends on the dirt tolerance of the co-inhabitants) and scoured with substances rude to the nose. In this world, crotch-sniffing is frowned upon, and even a surreptitious lick of an arm or mouth is only sometimes tolerated. The evocative and emphatic smells that people saturate their clothes with during the day — piles of smelly reminders of their presence — are mysteriously spun and heated away in large, noise-making machines, emerging as replacement "fresh" odors that call to mind nothing at all. And on very, very bad days the dogs are led to the fecal-urinous room, piled into a slippery-bottomed vat, wetted, and made to smell like taxi air freshener.

But oh! when it's time for The Walk: the olfactory scene changes. The great outdoors provides stimulation on the ground and on the breeze, on objects passed and passing objects. Each time the front door opens a new scene arrives, a scene of what has recently happened, what is now happening, and even a bit of what might be happening down the street. Not for lack of seeing, but clearly my dogs' journeys down the block are not made of visual landmarks. We launch outside a few steps above the sidewalk, and before we turn onto it the dogs' noses are in the slipstream of sidewalk currents, three feet above the ground, occasionally raised up high for a tall sniff. As we proceed, we pass a loamy tree pit, whose iron rail is messy with the passing dogs of the day; the sulphurous emissions from a work crew on the street, wafting from between parked cars; a frightening garage door that sometimes belches open with activity and rushes of air; the busy passage of a community of birds from apartment house to low tree branches, the berries dropped by the birds on the pavement, from mouth or ass; the edge of the corner building, where winds race uphill from points south; the slick-slippery feel of the marble stairs down, requiring more deliberate steps; the bench where a man sleeps at night, littered with his leavings and the stale, fetid odor of his clothes.

When I came to realize how central smelling can be to the dog, I began doing concerted "smell walks" with both of my dogs. On these walks we are not trying to make good time. Never do I pull them away from *that spot* they've been at for impossibly long; I celebrate rather than worry over their abiding interest in other dogs' rumps. We do not hurry to get around the block, to get home, to get anywhere but wherever their noses lead us. There is no place we need to get to, or time by which we need to get back, to satisfy smell-walkness. Instead, the walk is defined precisely by how long and how much my dogs can sniff in. Sometimes the walk involves a lot of not-walking: standing, nose buried into the earth; nostrils pivoting their heads around to find the dog who left *that message;* even lying down with nose high in the air.

Our dogs' smell-walk routes are almost never the routes — or the paces — of the walks that we humans concoct for them. As with many owners, when I head out on a walk, there is often a rectangularity to the episode: down, over, up, back. A walk in the human book has a time limit: "until I need to leave for work," "until he does his business," "until one of us is tired out." A walk in the dog's book is much more determined by circumstance: the course is irregular, doubling back and turning suddenly. On many days, there is no end in sight — until one suddenly arrives at it.

It occurred to me that, should I truly want to grasp the dog's sensory experience, it might make sense to begin with the dog's walk. I'm already along for the ride. As dog owners, we are implicit — if rather uninformed — accomplices as our dogs sniff their way down sidewalks, along road edges or park paths. How much more might it take to convincingly experience what they smell? After all, everyone who walks a dog surely sees, after a few goes around the block, the dog's heightened interest in fireplugs, a tree branch fallen overnight, or a building scaffold newly erected. Over many, many walks I began to get better at predicting what would be of interest to my various dogs, following their noses. *That fencepost looks recently moistened,* I found myself thinking. Or, *Oh, that leaf pile looks pretty appealing.* This is, of course, but a tiny minority of the dogs' smelling experience. Everything they (seem to) put their eyes to, they are nosing: the passersby, the wash of air from a car door opening, the ground around the park bench, the plastic bag being swelled up by a breeze, or the helicoptering maple seed being sent down by the trees. I *see* it. But do I dare to *smell* it?

"Do you ever smell what the *dog* is smelling?" Avery Gilbert first says when I ask him for a suggestion of how to become better at using my nose. "You want to get as close to the ground as you can get to smell things," Stuart Firestein tells me, "because that's where the molecules are." While I know this logically, I had not followed it through, to, as Gilbert tells me, "get *down there* and sniff at it."

Indeed, we don't generally *sniff* it, whatever *it* is. And so I determined to sniff at it. To smell my neighborhood as my dogs did.

I found myself surprisingly unworried about the obvious: that the reason we don't sniff what dogs are sniffing is, essentially, simple self-preservation and survival. There are funky, foul, downright fetid smells on the sidewalk. Our species may be the robust, successful species it is because we do not ingest the noningestibles down there. Instead, I considered the logistics: how could I get at the smells? To start, I figured, I must be willing to lose some of my restricting bipedal configuration. If indeed being upright, nose too far from the ground, is what keeps us from smelling, why, I'd go on all fours. When my dog lingers with nose planted at the base of a tree, so shall I plant my nose. If he gets intimate with a patch of invisible interest in the grass, so shall I pursue that interest. If he makes inquiries about the pant cuff of a passerby, so shall I. Passerby willing.

I began strongly. On a cool summer morning Finnegan, Upton, and I launched out off the stairs in front of our apartment building. Right away Finn angled for a tree guard: the impotent short iron fences that ring street trees in New York City. He examined the edge, which looked recently moistened, with the precision of a watchmaker diagnosing the innards of an ill watch. When he came up for air, I took over. I had to kneel awkwardly, one hand ringed with leashes and the other on the rail. I bent very, very close — too close. Emboldened by Finn's thorough vetting of the spot, I sniffed with abandon. A strong, bright smell hit me. I sniffed again. It was not urine, I thought with appreciation. It was, simply, the cool astringency of paint on metal. As I pulled back my head, my dogs stood to the side, watching me. A couple walking up the hill took a wide berth. I got up, suddenly self-conscious, brushed off my knees, and let Finn pull me away and down the sidewalk.

"No one questions a dog when they smell the environment," says Kate McLean, a self-described "multisensory" artist from the UK who has herself been pursuing urban odors. It was only too apparent to me and everyone else that I did not have the canine bona fides to get away with sniffing tree guards. After that brief foray into sidewalk sniffing, I lost my courage. To build some smelling bravado, I followed the brave — in this case, McLean, who bravely smells her way through cities with an interest in depicting the particular smell clouds to be found in each.

McLean was visiting New York and invited me along on a "smell-mapping project" in Williamsburg, Brooklyn. I found her, a bit flushed and with a black stingy-brim fedora atop her head, waiting on the sidewalk on a warm September evening. She is lean and fine-featured, with an easy smile. Her shoes were well worn, presumably having walked her nose along many miles. Umbrella in hand, a cheeky nod to the UK style of "tour leader," she addressed a cluster of two dozen smell-interested folk gathered around her. They included artists who worked in multimedia, photography, and with olfaction; memoirists and science writers; a couple of interested hangers-on; and two professional children. We were, McLean instructed, to take a tour of six or eight blocks and *simply smell,* recording what we noticed.

Well, it was not *so* simple.

"Be aware that there are many different kinds of smells," she cautioned. "There are episodic smells, which drift past in a moment — a person, a draft from smoking, a truck — brought on the breeze. Especially at street corners. So at corners, stand and turn around."

Apart from the wafting smells, there are also static smells that have been absorbed into materials. "So sniff walls, touch plants, go into stores," she instructed, watching smiles creep onto every face. A few people exchanged raised-eyebrow glances.

And furthermore, "smell voids" are just as important, she said. Given that we get acclimated to smells, we should be alert to when we are *not* smelling anything. McLean recommended using one's own smell as the natural resuscitator of one's sense of smell: by burrowing your nose in your own skin, the receptor cells have a moment's pause to recharge and resume their attention to the environmental odors.

McLean handed out handmade accordion-folded maps of our route, with "nose points" along the way to remind us to pause; to smell high, low, close, and deeply; to breathe in passing smells and constant smells. At each stop we were to record five scents. Five! I inhaled through my nose as we listened to her. I smelled nothing, but by the walk's end my handwritten notes would be overflowing the page, turning around corners and ducking under other notes much like the routes of the smells we were chasing.

McLean has traveled the globe smell-mapping, from Amsterdam to Pamplona; Glasgow to Newport, Rhode Island; Milan to Edinburgh; and Paris to Singapore. After each walk, she has translated the walkers' records into beautiful maps with a colored, topographic styling marking the sources and spread of odors. Washes of colored dots mark migrating smells. Each city, she suggests, has a background smell that specifies the place. In Amsterdam in the spring, it is the "sugary, powdery sweetness of waffles" and the water of the canals. And each city has distinguishing particulars: Edinburgh's map, for instance, includes fish and chips, malt spewing from breweries, and the scent of "boys' toilets in primary schools." Heaven knows how she knew that one.

On an earlier visit to New York, McLean mapped what she described as the city's "smelliest [square] blocks" — on the Lower East Side between Allen and Eldridge, south of Delancey. While this part of the city has a varied history — partly manufacturing, partly sordid — its recent sprouting of multimillion-dollar condominiums challenges her claim. Still, the final smell map included landmarks of sawdust, trash, car oil, and cabbage, as well as long swaths of dried fish and cheap perfume.

Geographer J. Douglas Porteous called olfactory landscapes like those McLean maps "smellscapes." Cities, it has been claimed, are identifiable by their scent. As a freshly baked baguette invokes contemporary Paris, the characteristic odor experience of a city may come from food or spices sold on the street, the marine air that fills the city's avenues, or the detritus of the teeming populace. Certainly there have been smellscapes, celebrated or not, for thousands of years: in ancient times, temple builders mixed milk and saffron into the plaster; mosques were built with musk and rose water worked into their mortar. On being rained on or warmed by the sun, the buildings effused fragrance. For many years there were regular weekly smells: the warm wet smells of washing, the hot iron on linen; the scent of "baking day."

The idea of smellscapes has caught on in the field of urban design, whose proponents have an eye to celebrating and improving the sensory experience of city residents. Some of the programs cities have enacted are pleasingly quirky. In the Netherlands, pedestrian plazas have been designed to include plants that may have relaxing, therapeutic effects. Since 2001, the Japanese Ministry of the Environment has supported the maintenance of "One Hundred Sites of Good Fragrance" countrywide: national landmarks as significant to the history and culture and life in the country as visual landmarks. They include "the smell of deer seen at Jinhua mountain"; a house smelling of the animal glue used to paint papier-mâché dolls; "one hundred thousand peach blossoms at a glance"; and, in the biggest city in Japan, "Kanda's streets of used bookstores."*

This turn into designing and celebrating smells follows a long history of complaining and worrying about smells in cities. And most of this cogitation was because cities smelled utterly, horrifyingly disgusting.

The streets of Manhattan were designed as a grid not only for ease of way-finding but also for ease of smell-letting. What this implies is that smells needed to be *let out.* Indeed, the curvy, narrow streets of old European cities — Paris and London, for instance — were widely seen as attracting and providing breeding ground for stinks. "Paris may be smelt five miles before you arrive," it was claimed; the atmosphere around the cities of Italy was saturated with garlic. The smells of nineteenth-century Paris were described as "intolerable"; the "monstrous" city of London was "strewn

* A recent topic of research for scientists who publish in the *Journal of Pulp and Paper Science* or who study "material degradomics," the "smell of old books" has been deconstructed and seen to overlap with vanilla, mushroom, nylon, and "grassy notes with a tang of acids." Since the smell is literally the degradation of paper, ink, binding, and glues, it is not obvious that Tokyo's bookstore blocks will smell the same in another century.

with excrement, mud, decomposing animals, meat, vegetables, and blood"; odiferous industries like tanneries and breweries were cheek by jowl with residential sections. Both London and Paris had episodes known as the Great Stink — both caused by a failure of the prevailing sewage treatment approach.*

The idea with the Manhattan grid, reaching from river to river, was that smells would swirl down the streets and out to sea. The New York commissioners reported that this would "promote the health of the city" and used words like *free* and *circulation* to vaunt their approach.

The grid did not save the city from its smells. And "smells" in this context were mostly "noxious, horrible smells." Consider the state of New York City in the mid-nineteenth century, when horses — then

*At the time (1850s) in London, cesspits under residential buildings filled with the wastewater and excreta of their inhabitants were transferred to the Thames and summarily dumped in it. Apparently this worked well enough, with the exception of a particularly dry June, when the river ran low and the wastewater ran high. The river "ripened" to "a peak of pungency" and lasted for two weeks, until the heat abated.

the city's main transport devices — shat freely and plentifully in the street and were often left where they fell and died; the contents of chamber pots were summarily dumped out of windows; and fear ran rampant that miasmas (foul gases emerging from the ground) were leading to a cholera epidemic. Conditions were sufficiently bad that a governmental "smelling committee" was officially called together, tasked with seeking out sources of ill smells by nose(s).

The notion of contamination by smell continues today, in small doses, with fear and retreat from people who smell bad, as though disease were conveyed by odor. Though long discredited, this anxiety had its root hundreds of years ago, when "de-odorization" projects began. In particular, streets and sidewalks were paved to tamp down the miasmic emanations; plaster covered walls that might seep with foul-smelling vapors; industries, reeking of paint solvent (vaguely banana) or cleanser (soapy), were moved away from residential areas.

Today's city has done away with the horse and chamber-pot leavings, but it still smells of the city's occupants and wares. Concentrate as you walk down the street and you'll notice the wash of odors out of open store doors, warmed inside and released in bursts by customers' entrances and exits. People are perfumed with bottled fragrances — or the bodily odors that bespeak a lack of fragrances. But one can walk many miles in an American city without being hit by any specific smells. One of the responses to the overwhelming array of noxious odors in cities has been to try to do away with odors altogether (as "*deodor*ization" implies) or to cover them. Ubiquitous chain stores homogenize cities, and there is a booming business of branded scented environments, in which retailers diffuse a fragrance intended to invoke "car showroom" or "fancy hotel." The result may be the waning of the characteristic odor of a city. Should Paris stop smelling like bakeries and Gauloises, Vancouver like the spray of sea salt, or midtown New York of hot garbage and food carts' emissions, some part of the place will disappear.

The British urban planner and designer Victoria Henshaw started thinking about characterizing — and perhaps commemorating — the remaining smellscapes. Building off the idea of city "soundwalks," in which urban explorers actively listen instead of just listening in the background, Henshaw thought to bring the notion to collecting smells, too: smelling actively, searchingly, and intentionally — instead of passively and accidentally. McLean has used the noses of others, in addition to her own, to take off from Henshaw's early, labor-intensive work.

On the street in Williamsburg, our group was slow to begin smelling. To rally us, McLean brought our attention to a handout she had prepared. Stay hydrated, it instructed. Find hidden corners. Be not embarrassed. Finally, "Sniffing in public is completely legal," she thought to add.

Apparently, the activity is sufficiently nonstandard that she has had to consider this.

We set out at a meandering pace. On the sidewalk, among the generally steadfast and quick-footed New York City pedestrians, our group had a vaguely disoriented look. I came to learn that this is characteristic of those who are sniffing the air: a faraway, unfocused look to the eyes, head cocked or raised, expression somewhere between "Did I leave the oven on?" and "I just remembered a dream I had last night. I was in a car, with no pants . . ." Passersby step around you.

At our first "nose point" we formed a scrum of confused loiterers, nosing around for something to zoom in on. Stepping to the curb I got my first waft: a nose-pinching, clean-but-not-clean smell. A warm, scrubbed sidewalk: chlorine at battle with filth. Across the street, a food truck gave off an unmistakable taco-shell/chips smell: fried corn and used oil. The evening was sufficiently warm, we realized, to be full of smells: just as hot foods have more odor than cold, summer days have more odor than winter ones. Warmth makes many substances airborne, volatile, wafting up to meet any sniffing noses.

Our group began to self-organize, alternating poses by curb and by building, sniffing, then scribbling down a report. By the next street we were all but choreographed in our movements, flocking in pairs to local landmarks — a tree, an outdoor bench, an exhaust fan — and pointing our noses in unison into the air. People themselves become clues as to where to sniff. A photographer who had traveled with McLean from the UK, Sam Vale, bent under a bench backed against a food shop. I followed his nose. Wheatgrass (source: juice shop) prevailed on the sidewalk. But at the altitude of the bench, the pith of a leaf mixed with a definite spring-onion smell coming out of the exhaust fan. "This smells amazing!" he said and smiled.

We were into it. I sniffed the base of a tree: urinous. People walking by brought a chaos of fragrant odors: hair products, lotions, perfumes. Fried food followed a person with a takeout bag. "Use your other senses to guide you," McLean reminded us. I touched and crushed a leaf (pleasingly fresh). I followed my ears to a dripping air-conditioning unit (dank basement) and to a clean towel spanking the air as it was shaken out (dryer sheets). Each new or different feature spotted on the sidewalk aroused interest. A construction site's temporary fencing yielded a peephole into the site (dust, caulk, and warm brick) and, on its surface, a place for posters (fresh paper, paste). Even sights I would ordinarily veer around I began veering toward. With but a moment's hesitation I dipped my head into the olfactory space above an open trash can. It smelled sweet, almost tangibly so. The remains of recently chewed mint gum cast upward to my nose. What was usually stomach-turning had turned into simply news of the street.

Virginia Woolf once tried her hand at smell-walking, in a manner of speaking, through her biography of Flush, Elizabeth Barrett Browning's cocker spaniel. He wanders off into Florence "to enjoy the rapture of smell" on the streets — "the rough, the smooth, the dark, the golden" — smelling brass being battered, bread being baked, hair combed, cloth beaten, and men spitting.

I thought of Flush locating the "swooning smells" of the gutter as we hovered around a sewer drain and tried to catch a spiral of air. Far from swooning, it walloped me. Was it salsa? It bit us with its sharpness. "Chinese food" was offered. The photographer, Vale, poised over the drain. His head cast around, neck extended, the very image of a newborn animal lurching for its mother. After a moment, he said, simply, "Garlic." Our group erupted in murmurs of appreciation. The question of whether or not we wanted to be smelling garlic from the drain was trumped by the great satisfaction of identifying it.

Night began to fall. We had been walking for hours. We were shooed away from a restaurant where, even at a distance, our interest in the outdoor diners' food was a little too conspicuous. At the walk's final odor-promising corner, I faced a plain bricked building, the spent neon sign in a high window signifying a particular kind of dark, stale-beer-smelling bar within it. Instead of stale beer, I caught a very pleasant savory smell. It was suspended in just one small invisible cloud of air, and I had to stand on tiptoes, head raised and nose up, in order to catch it. When I stepped off the curb, it was replaced by a cloud of something darker, gray and waxy. I took a quick look around for likely sources — perhaps someone was walking by with a plate of food — and found none. *Come over here; there are smells here waiting to be caught* I indicated to one of the very good smell-catchers. He beelined over, stood facing me on the curb's edge, and we sniffed. Gas? Tar? Not quite. Then he stepped off the curb in front of a vehicle parked on the street and inclined his head toward the grille — just about where you'd put your head if you wanted to be squarely concussed by a car. I went in for a sniff. Warm air greeted my face: the car, an SUV, had been recently parked. A smoky, waxy note curled out from the engine. It smelled delicious.

How did it come to smell delicious? When one begins actively searching for smells, one finds . . . not a lot. Just opening your mind (and nose) to the possibility of smells isn't enough to actually smell anything. And actively sniffing is an oddly tiring exercise. Try sniffing now, and keep it up for just thirty seconds. Done? Chances are, you quit about halfway there (or wanted to). And furthermore, you probably noticed not a single smell.

So catching a whiff at all can be exhilarating. But here our undercooked relationship with odors kicks in. In English, most words for smells are words for their sources. While the sommeliers and perfumers among us may have a vocabulary to describe that whiff, most of us need something more. To name it — to know it — we want to know where the smell comes from. If the name and the source are not the same, our work is to resolve them satisfyingly. McLean remembers someone worrying to her that Paris seemed to have a background note of honey. Why should the city, not a center of beekeeping, smell of it? She tracked it to its source: not to a hive nor a covetous Pooh Bear, but to the wax polish popular in the city rife with parquet floors.

If a smell wanders off before being pinned, a beetle to an entomologist's board, the frustration is stark. If the smell is traced to its source and its name, it feels truly caught, captured, collected. The car grille smelled delicious because it was, very clearly, hot oil on hot metal. I recognized the smell, but my certainty was buttressed by my recognition that I was, after all, sniffing into a recently used engine.

Buoyed, I stepped into the street, momentarily free of traffic, in search of the savory smell still at large. Catercorner to the bar across the broad intersection was a shop radiating light into the dusk. Its glass doors, trimmed in red and thrown open, faced the intersection. Aha! *This* was the source: a bakery. A terrifically obvious smell, now that I saw its source. The bakery was outputting a caramely, buttery smell, some of which was surviving being tossed this way and that by passing cars and weaving its way across the street to my nose as I stood on tip-toes. "I think I need to go there," said one of the other walkers, catching the drift and darting toward the light. I did not. I had my source, and the moment was complete.

At the walk's end, McLean changed senses on us. Pulling out a flip book of paint colors, she asked for the color we would use to describe one of our smells. I chose Pantone 1245C, an unlovely yellow-brown with a hint of green, for the first smell I pinned, the chlorine of the washed sidewalk. "Everyone always picks a yellow-green," she said. But another walker found the caramel smell to be maroon, and our shared wheatgrass was remembered in a light mint. Garlic and the tarmac were an ashen purple. Gilbert and others have found correspondences between color names and odors, even with smells that don't have clearly colored odor sources like lime or banana do. In one study, odor of civet — an artificial version of the odor from the anal sacs of civets, used in perfumery — was roundly considered "brown," and another perfume favorite, bergamot oil, was most often seen as "yellow."

I headed down into the nearby Bedford Avenue subway station to catch a train home. To my great surprise, even as I approached the entrance, I could smell it. Now, I'm sure that the subway always smelled — but never before had the smell reached out and poked my consciousness. I slowed on my descent down the stairs to the station. The smell of youth (wet, shampooed hair mixed with teenage body odors) and of decay (the moldering of the walls under accumulated dirt and water). I smiled.

When you notice the smell of decay and smile, either something is very wrong with you, or your relationship with smell has changed. The smell walk had begun to change us. Smells were to be noticed, collected, considered, not just avoided or spurned. The deep funk of the subway in the summer is plainly awful. But sometimes, awful smells are awful because of their incongruity: a classroom with an eggy smell, a restaurant with the smell of air freshener. Like the garlic and car grille, the stink of the subway suddenly felt *honest:* it was what one would expect the subway to smell like. What is truly off-putting are disingenuous smells: ones inconsistent with how a place ought to be. We have a sense of familiarity born of experience more than consciousness. We know the sounds of our home — and hence might notice if it is "too quiet" — and we know what a baseball game sounds like — and would be disturbed if it sounded like golf. We expect vision and taste to align, so we want our orange juice to be orange, not purple, and any food that is flavored deceptively (bubble gum chocolate) is foul.

So, too, with smell. In 2005, all of New York City was up in arms because of a sudden odor of maple syrup oozing over the city. While maple syrup might rank among the most preferred smells of many residents, the overarching response was terror — as the smell made no sense in our urban jungle. Later, once the weather pattern (a cool night in winter and a "lid" of warm air trapping odors caught on the wind near the ground) and source (probably a flavor manufacturer across the river in New Jersey) were identified, the smell could be enjoyed for itself.

> After a little time walking and thinking, Dominic decided to smell out Grandville, his usual practice in towns he was visiting for the first time. He raced up and down avenues and alleys, rubbed himself against various poles, lampposts, cornerstones, and trees, inquired about the population and the town's history — how many members of each species it contained, the birth rate, when the town was founded and by whom, and why — looked up the oldest landmarks, smelled them carefully, asked about the climate at different times of the year, learned what the salary of schoolteachers was, and the price of tangerines . . .
>
> — William Steig, *Dominic*

On the five-minute walk between the subway stop and where I rest my head, I caught curls of air from basement dryers, grilled-meat trucks, circular-sawed wood; urinous wafts of mysterious source; a single smoker exhaling as he passed; curry from an open window; and menthol from a jogger's sweaty legs. This, I realized, is more or less the smell of my own block. One psychological study that asked undergrads to find their way, blindfolded, in a room divided into grids of different scents, found that they could navigate using smell alone. Could I find my way home by smell?

My longtime and well-loved dog Pumpernickel once strayed from our home in a coastal town in California. I returned home late at night to find the front door wide open, light streaming out, and a house that was definitely too quiet. Pump had, apparently, opened the door (these were days before dead bolts were de rigueur) and walked out. I raced up and down the street calling for her. Stoppering my growing panic, I began to devise a plan of action, retracing all of our walks together. I phoned some friends who ran a dog-food shop in town to ask them to come by and stay at the house while I searched by car. Imagining where she would go, if entirely left to her own devices, was trying: I was bereft.

When my friends pulled up to the house twenty minutes later, Pump jumped out from the back of their pickup. Between jumpings of joy, I asked how they found her. They had driven by their store on the way over, they told me. Pump was sitting out front, awaiting opening hour.

Now, we had many times gone to the dog-food store on our walks. It was less than a mile away. But we had approached it from so many different angles: how, I wondered, had Pump gotten there? Had she turned right at the eucalyptus, headed seaward, then taken a left at the bagel shop? Did she follow the crow-flies path, cutting across backyards and going through back alleys? Did she smell her way?

Walking through my own block's smells, I considered her olfactory navigation. Though the smells appear vague to us most of the time, they are very particular — enough to be landmarks. Sailors use smell in navigation: "(A)n old salt is said to be able to smell fog, rain, wind and snow. In calm, fluky weather, especially near shore, the knowledgeable shellback can often sniff out a breeze to keep his vessel going. He does this in large measure by sorting out what he smells — dampness from off the sea tells of a sea breeze or a fog while the aroma of new-mown hay, clam flats, or a pigsty warns him before the first ripple is seen that the breeze will be off the land." The homing pigeon's gift to return home, over hundreds of miles, appears to be a result of overlapping sensory awareness — including (but not limited to) smell. So, too, for dogs. In the First World War dogs were used by Britain as messengers or liaisons between the front trenches and the home camp, finding their way through some combination of navigational techniques including, presumably, the general smell of the area and the smell of (what was standing in as) home.

I turn left at the smell of curry, hop up the stairs, and am home. It's only my first trip to the trenches and back.

5: Plain as the Nose on Your Face

You can't miss the Monell Chemical Senses Center building in West Philadelphia. It's the one with an enormous golden nose jutting out of the building by the front door. A daring move, even for a center of taste and smell research: the triangle of skin and cartilage that fronts our faces is not widely admired. I have come to Monell to speak with an olfactory scientist, but I pause at the nose for a while and just look at it. How strange the nose is. While the human face looks bereft without a nose, it looks fairly silly *with* one. With our species' affection for gazing into each other's eyes and our focus on kissing or stuffing food into our mouths, we nearly overlook the organ in between. Well, not entirely. We break our noses; we surgically fix our noses; we pick our noses; we powder or sunscreen them. The nose leads our face as we walk and fills our bedrooms with snores as we sleep.

But it is often unlovely and largely unloved. It is reduced to an upside-down 7 by a child's hand, and may never evolve beyond this rendering. It is the stepchild of the face — the overlooked space between more esteemed anatomical parts.

For a facial feature that is so conspicuous, so apparent that it prompts its own aphorism, the nose is surprisingly unfamiliar to us. We might gaze at a cute turned-up nose or a prominent bulbous nose. We might suffer a sinusitis that arrives with the pollens of spring. Surely we savor the rush of dinner smells that hits us when we arrive home, just as we bristle at the wash of sewage odor emanating from the so-called sanitation department's hub.

We have many fanciful — if moderately obscure — words for *nose* in English. The beak, nozzle, snoot, beezer, bill, gnomon, nib, snot-gall; occasionally rendered as a bowsprit, a vessel, a boko, a snooter. The snitch, trunk, index, horn, the spectacles-seat. What we do not have is much knowledge about what happens after we sniff with that scent-box. Most of us know little about the smells of our world, little about how the nose takes in odor, and even littler about the smell system of the human brain. Scientists, surprisingly, feel the same way. George

Preti, the researcher at Monell whose group studies primarily "human odors," found the territory relatively untrammeled: "They were probably the last thing that chemists tackled for some reason." Scientific knowledge is only nose-deep, another olfactory researcher, Dr. Leslie Vosshall, told me: "The basics are easy," she said, "and then the hard stuff [about how smell works], we have *no idea.*"

"Nobody knows about olfaction," Stuart Firestein consoles me, as I admit that I haven't thought much about smelling in all my forty-some years. His statement extends not just to the layperson, but to biologists, and even olfactory neuroscientists like himself. Human olfaction was one of the last sensory systems to receive the scrutiny of science, upstaged by vision — how we love and admire our eyes! — and even the bench team, hearing and taste. But it deserves a look. It is not a coincidence nor an accident that we, dogs, sharks, and voles have noses. Noses house special cells that allow us to smell — and smelling is an evolved strategy to find out more about the world.

Smelling dates back to the ancient and single-celled prokaryotes, enabling them to avoid things that are toxic and head toward things that are beneficial. Today almost all living creatures in air, sea, or on land can smell in some way.* Smelling is *chemosensation,* the detection of chemicals, and we live in a world of chemicals.† The nose, then, is for seeing them, and figuring out which to pursue and which to avoid. "The biological nose," Firestein has written, "is the best chemical detector on the face of the planet." At the same time, he admits the olfactory system "is probably, like everything else in evolution, a bit of a kludge job" — a few systems all plonked in or around the nose, doing more or less the same thing, though in slightly different ways.

Humans have an odor problem. It involves our use of smell (scanty when compared to many other animals). Primates, including the human primate, are considered microsmatic — what is (probably unfairly) called feeble scented — as contrasted with macrosmatic dogs.‡ It involves our cultural

*Notably, cetaceans are an exception: dolphins, for instance, do not have an olfactory bulb, though they have a "nose."

†It is worth noting, in this age of distinction between natural and man-made or artificial products, that the term *chemicals* includes it all: every product of nature is wrought of chemistry.

‡Even so, some nonhuman primates can detect

sensitivities about smell (given that artificially scented "air fresheners" are not considered ironic or aberrant). Smell is reliably the sense that people suggest they would be most willing to lose. But also, most fundamentally, our aversion comes from a deep misunderstanding of smell. We distrust smells. Invisible somethings find their way into our noses: it can feel horrifying at worst, peculiar at best. Though we will put all manner of steaming, dripping, oddly colored foodstuffs *into our mouths* with alacrity, we can feel embarrassed, alarmed, or disgusted by an odor showing up in our noses.

The invisibility of odors accounts for some of this reaction. We rarely search them out; we more often experience them happening *to* us, catching us unawares. And there is uncertainty about what smells actually are. Here's what they are: they are molecules, flowing about in the air. Since the world is

---

smaller amounts of fruit odors called aliphatic acetic esters than dogs can. This makes sense for primate frugivores — or at least fruit lovers — as contrasted with the historically carnivorous ancestors of the dog. (Lest any primates feel too proud, know that Asian elephants are also very good with their aliphatic odorants.)

molecular, they can come from just about anything — gases, liquids, or solids (which continuously release a haze of molecules into the air).* In particular, biologists specify that odors are "small, low-molecular weight organic molecules." These molecules also must be a certain amount *volatile* — able to evaporate into the air, be caught by a nose, and cause the sensory cells to hum and purr. When we smell something, we are really *ingesting* it, after a fashion: the molecule is being absorbed by the mucus layer of the nose. In this way smell is different than our other, less alarming senses: what we see (through light reflecting into our eyes) stays *out there;* what we hear (through vibrations drumming our ear canals) ends at the ear; what we feel does not burrow into our skin, but glances off it. But in smell a bit of the source itself comes

*With exceptions: metal does not have a smell, for instance. Just as the "smell of sun" is the smell of things warmed by the sun, not the smell of the sun, the "smell of metal" — an iron banister after touching it, a handful of pennies — is actually the smell of the metal interacting with our own sweat, not a smell of the metal per se. Indeed, the researchers who discovered this describe metal's smell as a "type of human body odor."

into our bodies.

Most adults see smells as incidental, and as binary — very good or very bad. "The principal axis of human odor perception," write neurobiologists struggling with our detachment, "remains odor pleasantness" — whether we like a smell or not. Not only the unwashed masses but also "the greatest poets in the world," wrote Virginia Woolf, "have smelt nothing but roses on the one hand, and dung on the other. The infinite gradations that lie between are unrecorded." Sigmund Freud came out on the "very bad" side in general, equating the diminishment of our olfactory powers with the rise of our rational minds. "The organic sublimation of the sense of smell," he claimed, "is a factor of civilization." Sights are information; smells are judged. *Smelly* never means anything but "stinking." We like those who smell like our own social group and distrust those who are "smelly." "Being odorous," writes Jim Drobnick in *The Smell Culture Reader,* "is tantamount to being odious."

None of us began life smelling so monochromatically. We are born smellers. In early development, nasal chemosensors emerge before any other sensory system. And we have been in contact with odors from the get-go: as fetuses, we are in a wash of fluid carrying the odors from our mother's food. By birth the olfactory nerve is intact: better to search for the smell of mother's nipple and the milk it promises. Tiny glands around the nipple send beacons to the baby's nose.

In our first hours and days, we are tiny but *macro*smatic animals, exploring by nose more than by sight. A newborn recognizes his parents by the cloud of fragrance coming off them; his vision is still too blurry to see them clearly. A child's security blanket, or any favored raggedy, one-eyed teddy bear, is so loved due to its smell. If washed, it is changed — and sometimes rejected. Children are ambivalent about what adults find to be clearly horrible smells: they must *learn* to hate the smell of spoiled milk and flatus. They don't know that "skunk" is a bad smell; that "flower" is terrific.

But now, reader, take a moment to consider: What have you smelled today? Chances are, nothing. And if anything, it was involuntary — the baked bread as you entered the house, the thoroughly-sodden-dog smell that you'd sealed up in your car after that trip to the lake yesterday. When I ask people what they have smelled today, I often get a lot of searching looks. By adulthood, we have mostly forgotten that smell developed as a means of *discovery.* Most animals — our ancestors among them — use smell purposefully: to sniff out potential mates, to find delicious and nutritious food, to notice predators before being noticed. We have not entirely neglected these important discoveries, but instead of smelling a person, we smell his shampoo; the food is not found in the wild, but cinnamon buns and pizza joints can be navigated to by nose; and while we might not smell danger, we know the acrid smell of smoke and the added smell (the rotten-eggy compound mercaptan) of natural gas.

To restore smell to its rightful place we must undertake a simple three-step process: first, notice the smell at all. We must sniff it in, let it settle into the warm lining of the nose and snuggle into a receptor cell. Second, we have to be able to distinguish the smell from other smells — to simply note and remember their differences. Finally, we want to name what it is, or locate its source.

To begin, we'd better be clear about our noses.

SCHNOZZOLA

Nosed animals come in terrific varieties: mollusks smell with their tentacles; male silk moths with their feathery antennae; the simple nematode worm detects chemicals via an opening near its front tip. The elephant's periscopic sniff is enabled by its trunk — which is also used to examine objects and to caress other elephants. The domestic pig's nose has expanded into a perfectly lovely implement for rooting around. Star-nosed moles have a spectacular, fleshy nose with twenty-two radiating appendages that function as tactile sense organs, doing no smelling at all. Semiaquatic animals like the water shrew blow air bubbles to trap scent and then re-inhale the bubbles to smell them. Leslie Vosshall and her colleagues at Rockefeller University discovered that the mosquito repellant DEET works on the insects' "nose" — receptors on its antennae. The repellant is a "molecular confusant," they write, that scrambles the message to the mosquito about a warm-blooded target nearby.

The world of nosed animals can be divided into those with hidden noses and those with flagrant noses. We are in the latter group, naturally. Among primates there are also two nose types: the curved noses (strepsirrhines) — think of the cute-faced lemur, with its surprised eyes and tufted ears — and the simple noses (haplorhines) of most primates, including humans. The curved noses have the wet, naked rhinariums of dogs and cats. Among the simple noses, we can be further divided into downward-nosed (hominoids and old world primates) or flat-nosed (new world primates), based on the direction of the nostrils. So we're a flagrant, simple, downward-nosed creature.

The human nose is, anatomically, a *soft* organ, layers of skin and muscle, braced only by cartilage and fat on the inside. The outside is rife with sebaceous glands, the inside lined with mucus. A squishy, unwieldy vessel, moist and oily.

And it *is* just a vessel, most of it. "In man," Isaac Asimov wrote, the nose is "primarily an air vent and has no exotic uses." (That might depend on what, exactly, one chooses to smell.) Notably, the exposed "projection in the midface region," as it is romantically referred to, is not actually the *smelling nose.* Just as with dogs, while odors are hurried into its dark depths, most of the visible nose is just a cavern and humidifying chamber en route to the treasure of olfactory tissue deep at the back.

The smelling part of the nose — the olfactory epithelium where odors are received and translated into neural signals that make the brain say *cake!* or *kimchi!* — is the very end of the cavern. In the depths of the human nose, at about the point where the outer nose flattens into forehead — the midpoint between the eyes — is a postage stamp–sized plot of epithelial tissue. "You can't reach it with your finger," Stuart Firestein cautions, as though I might try. Sitting across from me, he wears a look marrying perpetual amusement and skepticism. His whitish hair leans toward unruly, but never delivers on the threat. My fingers stay in my lap, but I squirm reflexively. I make a note to refrain from telling my six-year-old this bit of news.

Let's not forget the importance of that pyramidal, protuberant vent, though. If it's congested and the sinuses swollen, smelling is temporarily disabled: we often lose our sense of smell — and thus of taste — when we have a head cold. Food designed for astronauts must be highly spiced and flavorful, for they are permanently congested: the fluid in their heads does not drop nicely toward their toes with gravity. As it clogs them, it deprives them of the enjoyment of what were beloved foods back on terra firma.

The appearance of the outside of the human nose does not reflect its internal architecture. The large-nosed among us have no more tissue dedicated to smelling, relatively, than the tiny-nosed. In both, our olfactory postage stamps are a very small amount of the entire nose. Hence the reason that humans are generally considered poor-smelling: we have less space for olfactory cells, and thus less sensitivity to odors, than macrosmatic animals like dogs. "It's this little teeny-weeny nose," Firestein says. "And it's very, very tight up in there."

A nasal septum divides the nose up the middle, making two vestibules: the most perfunctory of waiting rooms en route to the back. Each vestibule is lined with special glands that produce up to two liters of mucus a day. This soda-bottleful of mucus helps moisten the air (good for both breathing and smelling) and helps protect against any large or irritating molecules flying up into the nasal tissue. For when you give a good sniff, air flows through the nose at twenty-seven liters per minute — "gale-force speeds."

LIKE DOG, LIKE HUMAN?

The human olfactory system is more developed than we usually admit. But is it near dog levels? Anatomically, the comparison is stark: our nose is smaller and our sniff is less complex. The human sniff is not unlike that of the ancient canid: a bellowsful, a crude, imprecise pull of air in and out. Unlike the modern dog, our sniff is long and slow: we take a second and a half to sniff even once, pulling in as much air as might fill a regulation-sized softball. We have half as many genes coding for olfactory cells, and more of ours are not functional. We have less space for smelling — only one to two square centimeters of epithelial tissue; our noses house hundreds of millions fewer olfactory receptors and half as many *kinds* of receptors. If there's less territory to make sense of an odorant, then even if it manages to land in the human nose, our sensation of it goes . . . nowhere. We may notice an odor but not be able to identify, locate, puzzle out, or even react to it, before it dissipates and we move on.

Architecturally, our noses are children's block towers next to dogs' modern architecture: made of similar stuff but in a much simpler, more brutalist formulation. While the human nose does hold turbinate bones like the dog's, there are only three small bones, and they don't bear nearly as much olfactory tissue. Turbinates in a human nose are like minimalist modern art: simple Miró figures compared to the dog nose's healthy branching tree. And alas, my friend, you lack the "olfactory recess," the deepest part of the dog's nose, partially segregated from the rest. This matters: there is no place in the human nose for air to pull up a chair, sit, and be repeatedly smelled. Some scientists have suggested that with the movement of human eyes forward to the front of the head, we lost the space for a nose alcove. As a result, we exhale any inhaled smell right back out, scrubbing it from the gentle embrace of the receptors. This accounts for the sometime success of our frantic effort to purge a foul smell by blowing our nose.

We entirely lack the vomeronasal organ that forms the dog's second smelling route: in humans, it is vestigial and disappears before we are born. All our VNO genes are pseudogenes, no longer functional, so we produce no cells, no receptors, and make no connection to the brain. We do not seem to detect pheromones at all. "Sadly," writes Tristram Wyatt, an expert in all things pheromonal, "there is no good evidence for a human pheromone to make the wearer irresistible to potential partners."*

Psychologically, we are also different than dogs. We will trust our eyes over our noses. If there's disagreement between the senses, vision wins. Cherry juice made to look green tastes to us like lime; color a white wine red and even enology students taste it as red. Not only are we largely ignoring the input through our snouts, we sniff so infrequently that only an unusually strong odor reaches our consciousness. Any attention we bring to odors as infants is quickly learned away: "Baby smells an odor, mother says nothing," Mary Roach quotes a researcher at Monell as telling her. And so

*This is not to say that we might not implicitly detect each other's biology in some other way: researchers are looking at "trace amine associated receptors" as perhaps involved in detecting the presence of bacteria. If they do so, then Firestein says, "The types of things that people would have said the pheromones are doing . . . judging the relative health of a potential mate or rival," could be done by these bacteria receptors.

baby ignores the next smell coming her way. Our brains develop around non-smell things; the dog's brain is developed around smells. What the dog ignores, by contrast, is all else apart from the smell he has his nose in. As anyone who has tried to pull his dog from a long investigative bout knows, the dog practices impressive attention. Fitting that the Latin root of *attention* means "to stretch," to direct the faculties toward. The dog's nose stretches in all ways.

Nor do we humans celebrate the intimacy of smell. Should someone be close enough that we can smell him, we find that *too* close. "Most people in Western cultures don't smell at social distances," George Preti tells me — referring to both people's bodies and their smelling habits. We seem to bathe at levels matching our sense of normal personal space; in the United States, it's an eighteen-inch buffer radiating from us. Notably, vision and hearing allow for interaction with someone in social space without any violations of personal propriety — we can both see and hear someone comfortably. But to smell them would be intrusive at best. When I ask Preti if he smells people often, he laughs and says: "I don't want to get smacked!"

Fundamentally, though, our noses work *in*

the same way as the dog's. As with the receptors in the dog nose, there is no one-to-one odor-and-receptor correspondence. There is no single "vanilla" receptor nor "smoldering cigar" receptor, despite the instant familiarity of each. We specialize in our olfactory sensitivities: some odors we can smell in impressively small amounts, such as the banana-y odor of amyl acetate in .01 parts per million; others need to be thousands of times more intense for us to notice them. We are great at detecting and recognizing coffee, which, like most food and drink, has hundreds of constituent parts, but we are utterly anosmic to plenty of other molecules. Some other animals smell carbon dioxide; we do not. For an especially puzzling case, look at carvone, in the class of naturally occurring chemicals called terpenoids. It comes in two identical forms that are mirror images of each other. One form smells like caraway seeds, the rye bread of an old Jewish deli; the other smells like spearmint gum. Our brains read the same molecule as entirely different smells. Any model of receptor processing tries to bring a rhyme or reason to why a molecule smells the way it does to us.

Our olfactory neurons function just as dog neurons do: they serve to transmit the message of an odor's arrival to the brain, which then scrambles around trying to figure out what in the world it is. That's when you "smell" it: when the brain registers there's something there. In some sense receptor cells "know" what the smell is — insofar as each will allow only specific molecular shapes to bind to it — but they don't really know. It is the brain that knows (or doesn't), and that swoons with the rush of a memory of hot chocolate after a long winter's day playing outside, or balks at a urine smell in the subway, source unseen.

Olfactory neurons themselves are pretty special. In all animals, these cells regenerate about every thirty days. You trade out old summer neurons, which may have conveyed the loveliness of the lavender garden and the rank odor of warmed manure to your brain, for new fall ones, ready for apples fermenting and coats being unmothballed. This fact is extraordinary. Aging usually means deterioration: all our senses dim through damage and cell loss. Our hearing diminishes over time as we damage our auditory cells by the fact of merely living (and listening to loud music through headphones, waiting for subway trains, and standing too close to the fireworks). In the course of a perfectly normal life we age into glasses, then reading glasses, then bifocals. But there is not an odor correlate to staring at the sun, turning the headphones volume to 11, or touching a scorching cast-iron pan. Unlike the neurons that allow you to see, hear, or touch, whose damage can lead to permanent loss, the nose keeps growing shiny new cells.

EVERY MAN TO HIS SMELL

Despite the qualitative and quantitative differences between dog and human noses, throughout my investigation of our snoots, I heard a surprising comment from various psychophysicists and neuroscientists. The human "schnozzola," as Stuart Firestein sometimes calls it, is "quite good," he says. Dr. Noam Sobel, neurobiologist at the Weizmann Institute in Israel, holds nothing back: in his papers he writes, variously, that humans have a "superb" or even "astonishingly good" sense of smell.

At first, these claims are puzzling. Every walk outdoors with my dogs seems evidence to the contrary. To see them suddenly stop, turn on a dime, and hightail it back five steps to nose something invisible on the curb is to see that my nose is an inferior model (perhaps, in this case, given what might be emanating from the curb, to my great satisfaction). If my nose were "astonishingly good," then I should, in theory, be able to experience some dogness by just remembering how to turn the thing on.

Not quite.

The neuroscientific research does indicate that, in essence, our olfactory equipment is reasonably good. What it overlooks, though, is what you and I know intuitively: how we use our noses. I knew that I was not smelling the way my dog was: she loved to loiter on invisible odor markings on every surface, in every breeze; I rarely bothered to sniff.

On the other hand, there is one convincing demonstration of the range of our noses: breakfast. How was breakfast? Did you taste it? If so, you have just confirmed your sublime sense of smell: taste is 80 percent smell. When we chew food, we are basically loosening odorant molecules from their tethers, warming them, and sending the odor-laden air backward in the mouth, where it is but a quick journey up the chimney of the throat to the nose. If in your childhood you ever experienced or witnessed the classic milk-out-the-nose result of getting the giggles in the cafeteria, you have experienced the short connection between the mouth and the nose. Smelled food needn't go all the way out; indeed, it only has to rise to the olfactory epithelium. As we exhale while eating, air from the lungs pushes past the back of the mouth, grabs some warmed food odors, and sends them up into the nose's backdoor. At least, if you're being polite and chewing with your mouth closed.

This top-secret route is called retronasal olfaction. Humans are *terrific* back-of-the-mouth smellers. The retronasal route is largely responsible for the fact that we experience food as having any flavor at all. While the taste buds impart experience — of sweet, sour, bitter, salty, or umami — these experiences will never add up to what we mean when we think *breakfast.* "The sense of flavor produced is a mirage," Gordon Shepherd, another neuroscientist, has written. "It appears to come from the mouth." Breakfast's deliciousness comes largely from the experience of its odor, as you can quickly confirm by plugging your nose while you take a bite. The feeling of the food — the crunch of the toast collapsing into a softness — is still all there. Indeed, the feeling of the food may be more present than you'd like: repeated chewing and toast suddenly feels gummy on the tongue, not an experience most are hoping for with their morning toast. Unplug your nose and the taste comes back in waves: a yeastiness, perhaps a caraway seed, the rich note of butter. Your nose did that!

You might experiment with an orange. Pick out a firm, well-colored specimen. A delicious, bright-smelling odor will greet you as your thumbnail punctures the rind: pith, zest, the effervescence of orange. Peel off a segment, slice it in two, and pop a half in your mouth. Let it rest on your tongue, but do not bite. You might feel its juiciness, sense that it might be sweet — but, notice, you cannot taste it. Now chew! The orange has returned to your senses. Pinch your nose and it disappears. Unpinch it and it's back to citrus central.

The reason it tastes so pleasing may be what the neuroscientists mean when they say we have great noses.

So it is that our orthonasal (through the nostrils, breathing in) olfaction is surpassed by our retronasal (through the back, breathing out) olfaction. If you've ever watched a dog eat, you have seen that the conditions seem to be reversed for the dog. Though your dog may delight in rolling in dead squirrel and may lick a passing dog rump with seeming pleasure, if you put something unappetizing in his bowl, he sniffs it and turns up his nose at it. Dogs use orthonasal for examination. If, on the other hand, the food passes muster, it is usually gulped down. Chances are that there is not much or indeed any retronasal olfactory experience: airflow in the nose hinders smells from rising up the very long route from the mouth. Nor is the food even in the mouth long enough to be smelled — let alone savored.

SMELL, MEMORY

Since the olfactory neurons reach into nose on one end and into brain on the other, a single synapse — the connection between neurons — separates the world of subway odors and overperfumed teenagers from our fragile central processing unit. In two synapses word of odor has traveled all the way to the cortex. "One of the reasons I work on olfaction is that it's a very shallow circuit," Firestein told me. "You can get from the outside world to cortical tissue in the brain in two synapses — *two synapses*! In the visual system you'd still be in the outer retina." The nose reaches the cortex lickety-split.

Once in, olfactory information cascades through the brain, knocking us into sensation and remembrance. This process — the creating of the olfactory experience — is not deeply understood, even in humans. Every smell researcher I talked to, though, held out hope for the answer to how the actual experience of smelling is formed. Firestein, who, like many olfactory researchers I spoke with, came to his research topic inadvertently, told me: "The promise in olfaction is that it is one of the systems in the brain where we really could get from an initial stimulus interaction" — from the odorant — "to some kind of a percept." Such a thing is within reach of the science of vision. When we spot a parent's face in a crowd, we now know that there are specific cells in the visual cortex that identify the vertical and horizontal lines on the face and others that recognize the fact of it being a person's face and not a balloon face — even before it is linked to memory and we can smile and say, "Dad!" The scientific knowledge of what happens past the olfactory bulb, by contrast, is a little scantier.

What happens upstream? It is not uncommon for an academic paper on some component of olfaction to say that an interesting-looking part of the system is "unknown." How the brain translates a pattern of neural firing into recognition of a scent is still a mystery. Even the mechanism of the very first step, the receptors, is still partly unconfirmed. "On all these different fronts we have no idea," Vosshall tells me. "We don't know what the odor space is, we don't know how the receptors capture that odor space, and we have no idea how the brain takes all of this information and makes a picture." Avery Gilbert puts it more bluntly: once you get past the bulb, "All bets are off. Nobody has a fricking clue."

What we do know is that once we are two synapses in, many regions of the brain then hear about the smell: this includes various other parts of the cortex, the amygdala, the hippocampus, and the cerebellum. These landmarks are clues to unpacking some of our experience of smells. First, noticing and responding to a smell can feel automatic, unmediated by reflection. There is a reason for this: olfactory information goes straight to the forebrain, missing the stopover point — the thalamus — where all other sensory systems arrive when entering the brain. Without air traffic control, the scent is flying in under the radar. Our reaction to a smell is often just as fast as the noticing of the smell. Second, olfaction is the quickest route into the amygdala, considered the emotional center of the brain. "The memories you get from olfaction are always emotional memories," Firestein confirms.

"You don't smell something and remember an equation, or a page of text. It's always Grandmother's house, somebody's closet, first day of school, an ex-lover."

Third, the hippocampus is involved: the seahorse–shaped part of the brain involved in the making of memories. Sitting in a too-large overstuffed chair in the dark of your grandmother's living room; coming upon a decaying animal in the woods; a new boy at school scooching in next to you on the bus after gym. As we begin to process these memories, a rush of odor sneaks in on the memory's coattails. Later, the odor itself sparks the whole scene.

Indeed, if smell has a good reputation with us, it is for its role in igniting memories long hidden from view. Scents cause a scene to shine suddenly — a sun blazing out from behind clouds, coloring the space before it. Often the memory is accessible *only* through the trigger of a smell: as a molecule wafts into your nose, you are transported into a childhood far, far away, and into the head of the child who inhabited that time and space. The limits of brain science on olfaction seem apt, somehow, given the limited access the conscious mind has to the memories preserved by unknown odors. These memories are not *Proustian,* as the term has begun to be used: they are not held in crystalline suspension, undistorted by subsequent experiences and learning. But they are evocative, calling forth, as often as not, a summary of an experience.

Ask anyone for their oldest smell memories and you get these characteristic, emotive responses.

My father:

Gene, who worked in Dad's [furniture] shop, he was the guy who varnished and worked with the woods and he always smelled of the stuff. That was the smell of Dad, a bear of a guy. And Mom's attic smelled like — was it mothballs? It had a strident smell . . . she was so tough on Dad.

My mother:

My grandmother would share a room with me when she visited, each of us sleeping in a twin bed. She would put this talcum powder on herself in the morning — *poof poof poof* — and it went everywhere. I did not like that.

The inside of Daddy's hat. Holding it before handing it to him, hanging it up . . .

A rooftop tar warmed by summer, the smell of your grandparents' attic, the brewery or river or tree grove on a childhood walking route, blackboard chalk and rubber cement, a struck pipe or a rolled cigarette, Play-Doh and suntan lotion, wet wool. A smell caught by an inhalation can be friction to matchstick, igniting the dormant memory. Let loose from the downy dandelion head of memories is a seemingly single soft moment with a thousand threads that spin away. Interestingly, these memories are rarely brought on by the foul and noxious odors that vex us in our ordinary days; instead, they are colored the nostalgic hue of childhood. I catch a sharp, bright smell like that of old tapes and am brought to the round, woody, tobacco smell of my father's desk in his study, the wide drawers opening to reveal packs of cigarettes, untold sundry desk supplies, and long pads of paper decorated in his scribbled hand — and there he is, sitting before me, large and smiling, ready to pause to greet a daughter poking her head in. It is not one moment but all such moments that I see: a childhood stitched together by drafts of smells.

BECOMING SMELLY

Do we all smell the same? The answer, you surely know if you have a functioning nose and have not been living in a cave, is clearly not. And this is true of both meanings of the question: *How do we smell — is it the same?* and *What do we smell (or taste) of the world — is it selfsame?* For the former, a trip on the subway at 8:30 on a weekday morning will serve to disabuse us of the notion that it is applied fragrances — perfumes, colognes, scented shampoos — that make us smell different from one another. If anything, fragrances unify us in smells, whereas our individual odor is unique.

And just as the philosopher might puzzle over whether we all see the same color that we call *red,* it is an open question whether we all smell the same odor that we call by a name — the *strawberry* or *spice* that accompanies the red. We all smell different odor scenes: exactly what you smell of the world is different from what the person standing right next to you smells. This is part biology and part autobiography. On the one hand, there is good evidence that every person's olfactory genome is slightly different, resulting in individual variations in what odors each of us can notice and attend to. Selective anosmia — smell-blindness — to particular scents can be inherited. Some people with a particular genetic construction cannot smell isovaleric acid, a component of body odor, at all. People differ in the threshold at which they can detect the presence of various scents by several orders of magnitude.

On the other hand, we each learn odor preferences and aversions, and even cultivate degrees of attention or inattention. While we both may say we see *red* and smell *strawberry,* your red may be brighter than mine. And my strawberry may smell sweeter — and may bring to mind the intertwined smells of tomato-plant stems and warmed, pocked strawberries on the vine on a Sunday afternoon in the garden by the house. (Boy, those were delicious nuggets of berries.)

There are better or worse smellers, of course. "Many people claim they don't have a good sense of smell —" Firestein begins, then implicates himself: "like me. I don't have a very good sense of smell. It's almost invariably not a neurological issue." Instead, "it's sinusitis, it's inflammation, it's allergies . . ." The conduit to the epithelium must be clear. "You can then say to them, 'Well, how is food for you? Do you like food?' That, they taste fine." Flavor is intact, so retronasal smell is intact. They have a sense of smell and do not know it.

But the majority of the experienced difference between individuals in their sense of smell is within their control. Every perfumer is made, not born. Even each detection dog is trained for years before being sent out to find explosives or bedbugs or pine martens or illegally imported guava fruit.

How much are you willing to push out your nose and sniff at the world?

6:
My Dog Made Me Smell It

It is humbling to read the words of Helen Keller on the topic of smell. Left without two of her senses, it is perhaps not terribly surprising that as a child she became exquisitely sensitive to her remaining senses, especially smell and touch. Still, her descriptions of experiencing what she calls the "elusive person-odor" sound as if she could be channeling a dog outright. From someone's simple exhale, she could read "the work they are engaged in [for] the odors of wood, iron, paint, and drugs cling to the garments of those that work in them. Thus I can distinguish the carpenter from the iron-worker, the artist from the mason or the chemist. When a person passes quickly from one place to another I get a scent impression of where he has been — the kitchen, the garden, or the sick-room."

Might *anyone* be able to notice the scent impressions evident to Keller — and any attentive canine?

There may be people who are born noses. But most of us are just born with noses. Plenty good ones, as we have seen. Apart from retronasal olfaction, the other bit of evidence of the considerableness of the human nose is more abstract than comes with breakfast orders. The recent coding of the human genome alerted us to the astounding fact that around one percent of our entire genome is devoted to coding for the olfactory receptors of the nose.

One percent! At first blush, that might not seem like a lot, but our time spent smelling is much less than one percent of our lives, between what we are remembering, scheming, seeing, wondering, daydreaming, feeling, swallowing, uttering, digesting, respiring, suggesting, moving, or thinking from moment to moment. And yet the genetic blueprint for you reserved a one-percent slice for smelling, spit-spot and ready for use.

Psychological research has provided plenty of examples of the inborn skill of the human nose. They are perfectly normal abilities, and yet they are striking. Mothers can pick out the smell of their newborn baby's shirt from among other newborns' shirts within two days of birth; and newborns, in turn, can distinguish their mom's smell from that of other moms — as well as recognize the amniotic fluid they swam in for nine months. These abilities are, evidently, innate. Nor do we lose them after infancy. Children can pick out the smell of their siblings from that of other kids of the same age — even after not seeing each other for two years; what is more, there is evidence they can identify their friends by smell. We recognize our own smells — our own body perfume, captured in our clothes, or our preferred applied perfumes, woven between strands of our hair. An experiment asking college students not to bathe or use soaps or fragrances for twenty-four hours while they wear a plain t-shirt found three-quarters then able to pick out their own t-shirt from among nine other similarly unwashed tees. We have the same ease with our partner's smelly t-shirt. When asked, we find it trivially easy to tell genders apart by smell alone.

Recognizing one's infant or partner makes a certain amount of biological sense, even if you've never inhaled the odor of a t-shirt in the hamper and thought, *Ah, great, that's my brother!* Indeed, given how common it is in the animal kingdom to recognize family members by their odors — everything from the paper wasp to the Belding's ground squirrel to the spotted hyena uses odor to recognize packmates and kin — it would be more surprising if we did not.

Happily, if oddly, researchers have provided sundry evidence that our olfactory prowess ranges a bit farther afield than family and friends alone. In one study, nearly 90 percent of owners asked to sniff a pair of blankets, each rolled on, slept in, and generally dribbled on by a dog (one their own), could tell which of two blankets had their own dog's odor. (Notably, the owners did not always find their dog's blanket the more pleasant-smelling of the two.) Just as the dog recognizes his person by smell, we can recognize our dog.

In another case, blindfolded subjects were able to reliably distinguish closely related strains of laboratory mice. They could tell them apart by the smell of their bodies, fecal pellets, or urine. When I questioned the experimenter, Avery Gilbert, then at Monell, about the motivation for this experimental design, he laughed. For many years researchers had been trying to fractionate mouse urine, he told me, "to find the active molecule that was the differentiator between the two strains. It's a huge project chemically; you have to have lab techs squeezing mouse bladders into test tubes . . . I was just — *Can we smell it?*"

They could.

The physicist Dr. Richard Feynman described feeling prompted by reading about the skills of bloodhounds to explore his own latent olfactory powers. Since dogs can identify items that had been briefly touched by a person, this is what Feynman tried: he asked his wife to handle a single book from their bookshelf while he was out of the room, then return it. Feynman picked it out. Later this became a party trick of his, matching three books to the three people who had touched them. "[N]othing *to* it!" he wrote. "It was easy. You just smell the books."

You just smell the books. "People's hands smell very different," he suggested. "All hands have a sort of moist smell" — and, of course, the smoker and the well-perfumed and the person who habitually jingles the coins in his pocket all wear their habits on their hands.

You just smell the books. That very evening, I set out to test some of these abilities with my unpracticed nose. Having no lab mice at hand, but having good availability of both dogs and books, I cobbled together a few home trials. My dogs both rest on a short blue sofa during the day. Lying rump to rump, they bisect the sofa but seem to choose sides randomly. So I asked my husband to monitor their resting poses and let me know when they awoke, stretched, and wandered out of the room in search of other diversions. Then with no evidence left of their presence but their scent, I entered to sniff my sofa.

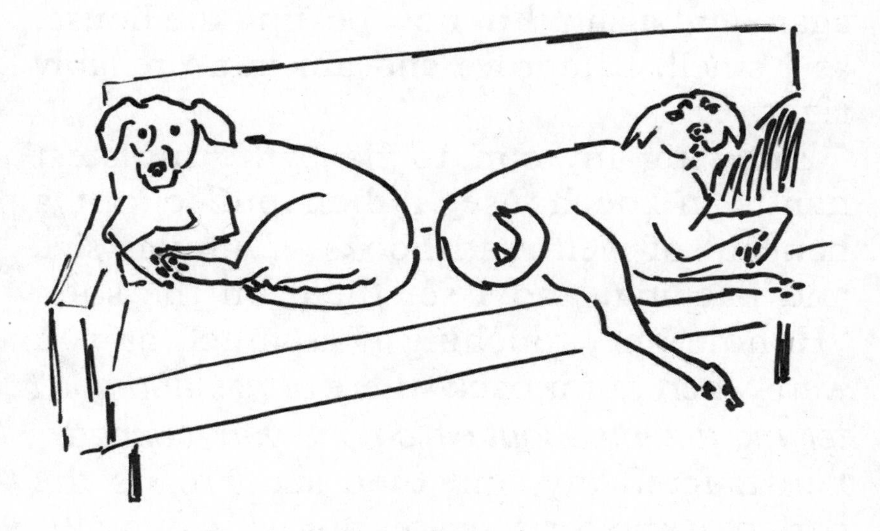

Belated apologies to anyone who has sat on that sofa. It does smell very much like the fur of dogs — not well known as a popular human perfume. I stuck my nose right into the cushions, smushing nose and cushion alike. The sofa faces north; its west end smelled clearly of Finnegan, musty and close and tangy. The east, Upton — a peculiar, somewhat unwholesome smell, recognizable also by it being not-Finnegan. My husband confirmed my choice. The dogs had returned to the room and were gazing at me dolefully. Upton took the opportunity of my kneeling posture to hump my leg.

Of course, I had a decent shot, one in two, of getting this in one go. So I have done it again and again. I'm now beating the house: as I smell, I improve and am more reliably right.

Knowing my son to have the moistest hands in the house, I then pulled out a handful of well-read books of similar size and backing, and I set them on the sofa. "Handle one, touching it as much as you want, then put it back where it was *and don't tell me a thing about which one you touched,*" I instructed my son, then six. This is the perfect game for a six-year-old, incidentally, should you have any around. Delighted, he did my bidding, an adult witness elbowed the books back in line, and I came in to Feynman the scene. I followed his very specific instructions — "You put each book up to your nose and sniff a few times" — and I was at a loss. Newish book covers are much more secretive about their identities than the pages they shelter. But one book did smell more . . . *warm-blooded,* maybe, than the others. I guessed.

Spot-on.

So clearly it's not that we *can't* smell; it's that we largely *don't.* And it's not only that we choose not to — although we do choose not to. One theory of the demotion of smell from our sensory lives is that our lost urge is traceable to our newfound bipedalism. When, in the course of human evolutionary events, our distant ancestors were equipped to stand upright, we not only began "flashing our genitals" to each other, as Freud worried; we also distanced our noses from the ground. The ground is a terrific source of smells — not only do things that emerge from the ground have an odor, most smells wind up on the ground, settling there after a trip on gossamer wings (or a current of air). Another, not-incompatible theory suggests that our sense of smell lost importance as our vision gained it, and thus began the shift on our faces and in our brains to visual primacy. As three-color vision appeared in primates' history, there was a decrease in the number of functional olfactory genes. Relatedly, a shorter snout allowed our eyes to develop closer to each other on the face, allowing for more overlap of our visual field and better binocular (three-dimensional) vision, which we cherish.

As I glance in the mirror, I'll admit, my nose doesn't look as up to the task as my dog's nose. Instead of projecting proudly forth like a dog nose, with nostrils open for all smelly comers, we house the nostrils at the bottom of the pyramid perched above our mouths. To be sure, it is the most prominent part of our face; indeed, the bones of our face are split in the middle to make room for it. Still, it hardly reaches out into the environment with bold and curious gestures. Instead, we tend to move our faces — noses with them — *away* from any approaching surface that might have a strong smell. One of the oddest things to do with your face is to push it right up into something: a person (often strongly frowned upon), the ground (only when we've been tackled), our food (we bring food delicately to the mouth — not vice versa — held on long silver implements to avoid even our hands touching it).

Based on the biological facts alone, the major determining factor in whether I smell an odor or not seems to be just this: whether I bother to try and smell it. Breathing normally about twelve to sixteen times a minute, that gives me twenty thousand opportunities each day, plus or minus, to smell something. We are, most of us, perfectly equipped. Sure, there is variation among people from the get-go, based on their distinctive genomes. But the major difference between me, on the one hand, and the sommelier and the perfumer, on the other, is one thing: time spent sniffing. Committed sniffing. *Practice.* Time spent putting our noses in things: under the horizon of a wide wineglass, over a small brown bottle containing a chemical formulation. Is there something in your hand now? Draw it to your nose and smell it. Likely, there's an odor there, but to trace that line from hand to nose — or even, heavens, to bend to smell the roses — is a bit out of fashion.

Practice is more familiarly applied to piano-playing, rifle-shooting, or tightrope-walking, but the upshot is the same for smelling. All these skills, in fact, actually depend on *sensory* practice: adjusting one's fingers according to what one hears; coordinating pressure on the trigger with sight of the target; vision and proprioception — the sense of the position of one's body in space — as one toes the tightrope. If you are a baker, you come to know when the apple cobbler is finished cooking. This ability comes from baking many apple cobblers, to be sure — and also from noting the intensity of smell that accompanies a cooked one. William James tells of a woman who could sort the cleaned linen of the residents at the Hartford Asylum by smell, from simple repetition of the daily task, and a man who came to be able to taste the difference between the bottom and top halves of a bottle of Madeira. That was a man who had tasted a lot of Madeira.

What Helen Keller had was *obligatory* practice. Deprived of hearing and vision, she was obliged to learn about the world with her remaining senses, and smell came to the foreground. But with some voluntary practice, could we learn to tell the iron-worker from the chemist? Or learn to smell when our children are home, a friend's illness, if someone else has been kissing our spouse, or how long ago a dog had been in a now-empty house?

Experiments have tested the added value of time spent practicing smelling. One study found people capable of learning the difference between two pairs of nearly identical molecules — one pair "green"ish-smelling, the other "oily-vinous." They merely sniffed the bottles repeatedly. Oh, and they were given an electric shock with one of each of the pairs.

That might have done it. Motivation might be required, but it needn't be aversive. Have we learned nothing from the years of miserable results of aversive dog training? Punishment does not lead to efficient learning; reward does.

And so another experiment was founded on chocolate, not electric shocks. First, a ten-meter line of twine was buried in a grass field in Berkeley, California, forming a simple trail with two straight segments connected by an angle. The trail smelled of chocolate: the twine had been soaked in diluted chocolate essential oil. Untrained volunteer subjects donned opaque goggles, earmuffs, work gloves, elbow pads, and knee pads to reduce their sensory experience to the olfactory. They were set on their hands and knees on the field three meters from the trail; each wore an airflow monitor below their nostrils to record their sniffing. It is not clear whether the experimenter said, *Go find it!* as one might to a working dog. But twenty-one subjects — two-thirds of the group — did successfully find the trail and follow it to its end within the ten minutes they were given.

Aerial video of one of the trials, played at four times actual speed, shows a definite sniffing creature, if not quite a doglike creature. Hands planted wide as she crawls along, the well-padded and sensorially restricted subject finds the trail without too much trouble, turns efficiently onto it, and zigzags her way down the rest in less than a minute. Her head veers back and forth between her hands, as one might vacuum a hallway rug absentmindedly.

Some of the subjects then agreed to practice the task: they sniffed along similar versions of the trail three times a day for three days. All of them found and followed the track each time — and they got faster with practice. By the end they were twice as fast as when they'd begun, speedily sniffing to the trail's end. They deviated less from side to side along the way. And they *sniffed* faster: going from one sniff every three seconds to two sniffs in that same time. Through various manipulations that must have required subjects with particularly good humor, the researchers also discovered that sniffing with both nostrils was better than with just one, and that it was the fact that each nostril gulps a slightly different square centimeter of air that made the difference.

Now, these were no dogs. All the humans were generally pretty laggardly, taking minutes to crawl ten meters — approximately the pace of an infant who has not quite learned crawling. The odor source was still lying there on the line in the grass — quite unlike a trail left by a source who then flees. And the trail was a simple, continuous path, which varied little in the practice rounds.

Let's not forget that these subjects were asked to smell for chocolate, not bottom-of-criminal's-shoe. No tracking dog is at any risk of losing his employment to these subjects. But the simple but significant result was: they got better at it. Practice made reliable chocolate-trackers of them.

"The Lost Muscles of the Nose"

So I resolved to practice. To simply sniff — and sniff again with the other nostril. To, first, bother to smell the objects that I touch and the air that I stride through. As I go through my days, to remember to inhale through my nose with the same intent with which I gaze with my eyes.

And also to sniff more assiduously, to practice in the manner of the chocolate trackers or the wine and perfume experts — expertise wrought of training, not genetics.

Knowing about the role of the humble sniff in dog olfaction proved practicable: if I was going to try to learn how to smell well, I'd better get my muscles in shape. My nose muscles.

You would be forgiven for not attending to your nasal musculature; indeed, one might not even realize that one *has* muscles in the nose. Editors of the redoubtable *Gray's Anatomy* apparently paid so little attention to the nose that of the seven "intrinsic," or interior, muscles listed in their 1901 edition, three mysteriously vanished by the 1989 edition.* This excising didn't have to do with a rapid devolution of the human nose, nor with exciting developments in our understanding of facial anatomy; instead, it may simply have represented neglect. Noting the muscles' absence from medical school anatomy textbooks, some physicians concerned with what happens to nose muscles after rhinoplasty (plastic surgery on the nose) confirmed in 1996 that those muscles still exist in our schnozzes. Not only that, these small, neglected muscles are critical for the nose to do its nosely duty.

A dozen muscles are dedicated to movement of the human nose. Each has an important expressive or expansive task.

*The lost muscles have since been found and are back in the current *Gray's Anatomy.*

There are muscles responsible for helping you wrinkle your nose with disdain or disgust; for the contracting and dilating that leads to, in the physiologist's terms, "a marked appearance of the countenance" and a look of contempt (that's the *Levator labii superioris,* should you need working on your contempt face); for changing the size of the nostril opening for breathing, working hard against atmospheric pressure; for enabling an expression of anger (think cartoon fire-snorting dragons and bone-tired racehorses); and, finally, for sniffing in odors.

Some of the sniffing work is done by "alar flaring": the energetic widening of the nostril — pulling out the wings — with the intrinsic muscles. In some people (sometimes those who go in for the above mentioned rhinoplasty) the nose wings are always flared, giving an appearance of moderate excitement or being perpetually mid-sniff. One study found that 40 percent of subjects, including a majority over the age of seventy, could not flare when asked to. Try to compensate with other muscles of the face, and the result is a combination grimace/snarl. (Which, temperamentally, might be expressive of the experience of *not being able to use the nose properly.*)

As it turned out, I'm in the 60 percent, and no workouts were needed, merely the knowledge of what I can do with a sniff. Just an awareness of the crosshatched muscles running along my nose made focusing on using them easier. Even the most inexperienced sniffers can voluntarily change how much air is taken in, how fast they take it in, how long the sniff lasts, and how many sniffs they take. Your brain, meanwhile, is not unimportant in this moment. It is the brain that determines what the smell is, not the nose. It takes into account the strength of the sniff: if you strain a nostril muscle sniffing hard, then the perceived intensity of the odor is less — you needed to work to get it upnose. In fact, just sniffing at all is sometimes enough to trick your brain into thinking you are smelling something: if you sniff and there is only pure, clean, odor-free air (inasmuch as this is possible), the primary olfactory cortex will probably register an odor. Both the flow of air and the activity of the muscles contribute to the brain's miscue.

Practice

It didn't start well. My first training at smelling led me to bouts of nausea, repugnance, and fear.

The solicitation was blunt. "The study involves smelling odors and evaluating them. The first visit will take about two hours." An olfaction lab at Rockefeller University run by Leslie Vosshall was looking for subjects. The call for participants was not in and of itself tempting. And yet there was an implied evaluation of the smeller herself that appealed to my competitive spirit: "There is the possibility that you may be invited back for up to ten visits in a relatively short period of time to evaluate more odors." This seemed the perfect opportunity to practice sniffing: regular bouts of compulsory smelling, in a school environment, no less.

I neglected to consider what ten visits at two hours a pop might be like.

The eligibility requirements are:

Age: over 18 under 50. Nonsmoker.

No allergies to odors or fragrances. No asthma.

No history of nasal illness or surgery that would affect your sense of smell.

No currently active seasonal allergies or upper respiratory infection.

Not taking medication that would affect your sense of smell.

I was eligible. I wrote to the study coordinator.

January 22. It's snowing in the morning as I make my way across town to Rockefeller's campus, perched on the far eastern side of Manhattan, behind a set of gates and with no crossroads inviting passers-through. Traffic slows and the city is partially silenced. The piling snow busily buries any city smells. With the wind, sensation is reduced to the tactile — cold prick points of snow pummeling my face.

At the university, I work to pull open the doors, extra heavy and nearly vacuum sealed. Inside, they've bottled the waiting room's perfume: part hand sanitizer, part wet newspaper and magazine. I am given an informed-consent form to sign that explains, gently, that 20 percent of subjects will have a temporary diminishment in their sense of smell after participating. Taking in the distinctive, acrid smell of Purell, I wonder if that would be all bad.

In the experimental room, three women sit stoically in front of computer screens, each with a large bin of small bottles at her side, and tap occasionally on keyboards. They do not remark on my entry. One woman rummages among the bottles, unscrews the cap of one, brings it to her nose, and unemotionally keys in some response.

One might take them for professional sniffers. But when I joined them it became clear that they were just hundreds of sniffs further into the game than I.

Peggy Hempstead, wearing a sweater and warm smile, greets me. She is the study coordinator and walks all the subjects through what kind of smelling adventure they are about to embark on. She brings a wide tray of tiny bottles over to me. I feel a shiver of excitement. One hundred white-capped brown glass bottles jostle and tinkle against one another as she sets the tray down. I will be asked to evaluate the odor of the contents of each bottle: first, its strength, pleasantness, and familiarity, but also its characteristics, its allusions, its basic "notes." Does it smell like fruit? A bit fishy? Or is it grassy? Urinous? Edible? They provide seventeen options. I hardly feel prepared.

Hempstead puts on a grave expression. "Some of the smells are *very* strong," she warns. She recommends taking a slow, serpentine route from uncapping to nose.

I listen but begin recklessly. Peering inside the first bottle, I see a few drops of liquid resting at its bottom. The merest hint of a liquid. In some bottles there will be enough to pool at the bottom, a tiny puddle I will swirl around like a wine snob to try to get more of a particularly weak scent. In others the liquid has crystallized and begun to climb the side of the bottle, lunging toward escape.

I open my nostrils wide for the first bottle and sniff until I'm full of air. I have no idea what I am perceiving. Clearly, this *smells,* I think, as I raise this, then the second, and the next bottles to meet my nose. But what *is* the smell? I rate each on "fishiness" and "acidity," but my assignments seem haphazard, random; I recognize nothing.

A dozen bottles in, I am reassured: a familiar scent. Coconut. The full, sweet roundness of it wallops me as the bottle is on its way northward. My childhood birthday cake; the macaroons my father coveted (and grandmother indulged); the pleasant, sweet nuttiness alongside a warm curry. Swimming in memories, I am momentarily victorious.

I am also about eighty-eight scents from finishing that day.

By bottle thirty I am disoriented, dizzied by the odors strobe-lighting my olfactory bulbs. There are innumerable odd smells that take me down long hallways with no clues where I am going: a deep sour note turns sharp — *Or is it spicy? Is it the smell of a person?* — and then disappears behind an unseen door. My head swirls with echoes of smells, but nearly all are utterly unplaceable — like a tip-of-the-tongue phenomenon, with no chance to retrieve the lost word. A *tip-of-the-nose* phenomenon. In this verbal purgatory one can judge whether an odor is familiar, and even name similar odors, but can come up with nothing about the name of the source. Some odors ring a bell, but more often than not it is a distant, muffled bell covered with blankets and smothered by traffic noise.

There are a very few glimmers of light. In the middle of a long streak of utterly unidentifiable smells, a particularly weak smell penetrates my brain and comes out with arms up in celebration: *Hello! I'm pencil shavings!* I sniff it again, redundantly, out of pure satisfaction.

Along the way, I also meet and identify bubble gum and almond extract. Another arrives at my nose with a *whomp* of recognition: it is my local health food store. I can almost hear the bells that announce my opening its front door — which simultaneously brings this distinctive odor. The next week I search out the owner in his shop. "Did you ever notice that there's a distinct smell . . ." I begin with some hesitation. People might not want to hear about their "distinct smells."

"Wheatgrass!" he cuts me off. "It's wheatgrass. Everyone says I smell like it."

One scent is unthinkably horrible. Worse, even as I suppress my own reflexive retching and jerk the bottle away, I know I'll have to smell it again — for the sake of Science. There is little that prepares one for mustering the courage to voluntarily bring a sweaty, vomitous wash of air close to one's face.

After some coughing, furious nose-blowing (as though I could force the odor molecules out of my cells), and a sip of blissfully odorless water, I get a reprieve with the next scent: *Smarties.* I loiter on it, letting the sweet, chalky pleasantness of it blunt my recent foul memory.

As I cap the final bottle, another scent suddenly fills the room: a sharp, sweet grassiness. I look around. Clearly, someone has spilled a bottle and released its smell into the room. But not only can I not see the errant bottle, no one else is looking around for the source of the odor. Hempstead, the coordinator, is walking calmly to a back room — not at all in the hurried way one might to escape an olfactory catastrophe. I think to call after her, *Hey! There's a smell!* and catch myself. This is clearly an odor coming from my own head. An olfactory hallucination — or maybe an aftereffect. I wait until Hempstead returns from the next room. "Do people experience, um, *imaginary* smells . . . ?" I ask. "Oh, yes!" She sounds delighted. "All the time! Smell echoes, even."

She predicts my future. It takes me two hundred minutes to get through those one hundred bottles — and that includes ones holding odors that I cannot detect, even with repeated and strenuous sniffing. I stumble out of the room. The entire episode has been stultifying. After I have put the final bottle back in its place and walked outside into the snowstorm, I search for the smell of the city. The only note that comes through is the exhaust from the idling cars, trucks, and buses stuck in snowstorm traffic. But an hour or two later, back inside, the scent echoes begin: mostly sweet, chemical-y smells coming from out of nowhere to suddenly punch my nose. Foods still have a smell — I am not one of the 20 percent who experience temporary anosmia — but they often have an *additional* odor: blueberries have an overlay of something rotting. Plain crackers are nauseating. Not only did my nose seem truly mediocre, unable to translate the odors into smells, it is now delivering smell when there is no odor. The afternoon is dotted with these smell apparitions until, finally, sleep blunts them all.

The next day I receive mixed news. "I am delighted to let you know that you successfully completed the screening visit for the smell study," the email from Peggy Hempstead reads. The good news: despite my impression, I can actually smell worth a damn. The bad news: there are nine hundred more smells to go.

Over the following months, I begin to catch on. I develop a method of *really sniffing:* inhaling as if my life depends on it. In my most ardent sniffs, I both pinch the nares in toward each other and flare them below, forming a belted skirt that rises in a breeze. Various nasal gymnastics — elevate, dilate, compress, repeat — become second nature.

As it turns out, there are many ways to sniff. And as with the dogs, the method of transporting odorant molecules into the nose turns out to be nontrivial. Without sniffing, only 5 to 10 percent of the air we inhale through our nose ever makes it to the epithelium. Unless we sniff, we do not smell. Olfaction is an "active process," says one academic paper on sniffing. Without any air flowing through the nasal cavity, the "olfactory scene" in front of us vanishes.

Indeed, by making careful recordings of the turbulent air movement around the upper lip, researchers found that we begin with a moderate sniff. Within fifty milliseconds, though, we adjust our so-called sniff vigor — a gauge of the force of our nostril pull and how much air we vacuum in — based on what we have detected. Given the speed of this adjustment, it appears that we are processing the odor subcortically, without awareness, and looping back to the nostrils to tell them to sniff harder or let up. Should the odor be very weak, or very lovely, we change the length of our sniff from sniff to sniiiiiiiffffff, increasing the force of the sniff and keeping at it. If it's a very strong odor, or very unpleasant, we sniff less vigorously and less long.*

We usually sniff many times, although one good sniff would be sufficient to get the odor in the nose. One "good" sniff would be a second or so long, taking in two cup-

*Even in sleep, our brain registers smell — and knows whether it is a pleasant odor or not (sniffing more deeply when it is pleasant). The odor tends not to wake us, though, unless it tickles what is known as the trigeminal nerve.

fuls of air. And when looking to deconstruct an odorant mixture into its components, two sniffs are much better than one.

Once in a while I pause and take in the style of the experiment's other subjects. Nearly everyone in the study seems to sniff one nostril at a time, matter-of-factly delivering the bottle first to one, then the other. As it turns out, our nostrils do different work for us: we alternate being left- or right-nostriled. One of your nostrils is a low-flow nostril. It is not worse; it just works harder or longer. So each nostril gets a slightly different view of the odor. Dr. Rachel Herz, a psychologist who specializes in olfaction, has shown that right-nostril sniffing seems to be strongly linked with "hedonic" responses. We perceive a neutral smell as more pleasant when right-nostril sniffed. People also are better able to discriminate unknown, new odors when using the right nostril than when using the left. The left nostril, leading ipsilaterally back to the more verbal left hemisphere, is what enables us to name the smell, once it becomes familiar.

Most of the time I feel smell-bereft. Many bottles, though housing something viscous or liquid, have no odor to my nose at all. I begin wondering about a prosthetic super-nose, able not just to leap buildings in a single bound, but to smell each resident inside as it did. The closest our society has come to inventing such a contraption came in the late-nineteenth century, when one Hendrik Zwaardemaker invented the olfactometer — a device that has never taken off. It consisted of two porcelain tubes, one of which could telescope out within the other. The user was to stick out from his nose, playing elephant. A nose-shaped fitting made the business end, and a handle underneath allowed it to be wielded like a radar gun. Rather than being a magnifier, it was intended to measure how much of an odorant was necessary to be detectable: the outer tube was saturated with the odor. But one could imagine it consolidating the smell of the thing it is pointed at, a reverse megaphone enabling odors to shout at us.*

*A contemporary olfactometer, the Nasal Ranger is a kind of demagnifier, allowing precision dilutions of an odor until it passes below the threshold for detection (thus permitting determinations of just how toxic a very stinky industrial odor is, for instance). Electronic noses, promising to do all the work of the dog's nose without the dog, are regularly developed but have not achieved near-dogness.

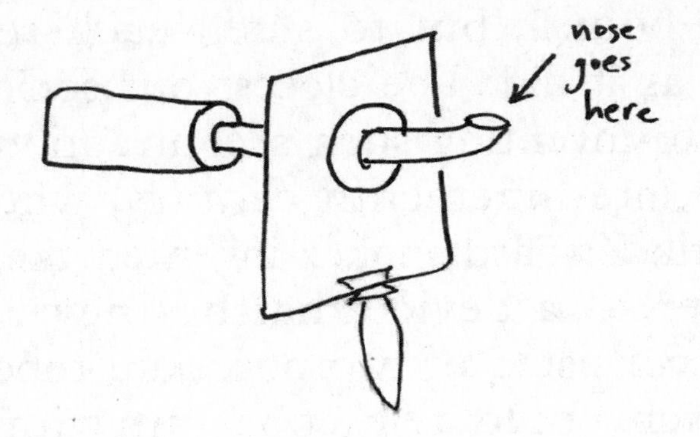

Identifying a smell was the highlight of any visit. Identification, though, was hard: the bane of my experience. I was missing words for smells, and the smells themselves were slippery creatures, eluding my attempts to pin them down with name or memory. Where was my language for smells?

Odors can be smelled, certainly, without the odor having a name. But our species — verbal and language-centered — has a hard time fully perceiving without words. The more words for smells we have, the better we will be able to see them.

But as it happens, most humans are lacking in two ways. First, most languages have very few words for smells. English has a countable number — countable on two hands. (Being generous, one might include *alliaceous, barny, farinaceous, fusty, hircine, mephitic* . . . and I've got four fingers left.) No "pink," "loud," or "rough" equivalents exist, no basic words for fundamental smells — or for lovely or dreadful smells. Most of the time we are describing the source of the odor: oh, that smells *like* coffee; it smells *of* cow dung. Or the source might be even more abstract: it might smell like "early summer," but what the *thing* that smells that way in the summer is — the honeysuckle, the cut grass, the damp earth prodded by tiny, forceful plant shoots trying to come through — is unknown. Instead, our words are more evaluative — the lovely or disgusting — or abstract — "green," "fetid."

Notably, not all languages are so smell-word-poor. Researchers describe the olfactory richness of the language of two Malay hunter-gatherer communities, the Maniq and the Jahai. Among the Maniq, whose culture values olfactory knowledge — about dangers, for use in foraging, in medicine — everyday language has dedicated words for scent. Odor words represent classes of perceptions, much in the way we use color words: as our "blue" covers the sky, the sea, and my son's eyes. For the Maniq, a word — such as "mi? huhu??" — might be used in reference to a whole category of similarly smelling things and experiences: snakes, soil, mushrooms, sweat, or walking in the forest. The Jahai name odors as easily as we name colors. There are a dozen basic odor terms, including one encapsulating "bat droppings, smoke, gingerroot, and petroleum."

By contrast, most of us (not living on the Malay Peninsula) are simply not trained, schooled, or accustomed to naming or identifying odors. Noam Sobel, one of the parties responsible for the chocolate-trained Berkeley subjects, describes putting a jar of peanut butter under a family member's nose while she closed her eyes. This person, he notes, eats peanut butter "every day." The result? Though "neurologically intact," she was unable to identify the smell. And this is a person who lives with a smell scientist. Research has found that when they encounter odors used or applied every day out of context, people are unable to name about half of them.

To be fair to humans, it has not seemed to matter to most of us whether we could wax eloquent about what we smelled, evolutionarily. A smell might lead us to flee, mate, or eat (or avoid those things). Not to speak. We still seem to have that thirty-million-year-old tendency to lead with an emotional reaction first, and search for the word to plug in second.

■ ■ ■ ■

Even when I cannot put a single name to the odor in one of those bottles, the satisfaction of recognition matches the great sensory pleasures of life: the *thonk* of an old refrigerator door closing; the smoothness of a ball bearing in a hand; goggling at the fitted intricacy and rhythm of a watch's gears. Over many visits I identify crayon; violet; something peppery that makes me sneeze for three minutes; flowers that have been left in the vase too long; nail polish remover; VapoRub; Red Hots; rum; apple lollipop; cooking meat; icing; Band-Aid; a biology classroom; Play-Doh; springtime; grass on a sweaty day; unspecified "candy"; charcoal; new vinyl.

I make anatomical discoveries. My sneezing turns out to be controlled by an accessory part of the olfactory system. The nerve that senses touch to the face and the top of the head, the trigeminal nerve, notes when irritants of various sorts are inside the nose, mouth, or eyes. We experience all manner of sensation — "cooling, numbness, tingling, itching, burning, and stinging" — as something we are smelling, though we are not. Spicy foods can be *felt* as hot and stinging; the trigeminal nerve detects pollutants in the air, prompting the nose to tingle and the eyes to water.

In one protracted sniffing bout, I discover the magic of *adaptation.* When one keeps smelling the same odor, the olfactory receptors, best at noticing something new, get weary and the neurons stop firing. This is a physiological phenomenon with daily application: enter your local coffee shop and, if you are a coffee lover, the smell of beans recently pulverized by expensive burr grinders hits you at once. A few minutes later you may still be in line, but the smell is mysteriously gone. The coffee shop has not changed; the receptors that noticed its odor are just momentarily quiet. You can awaken them by leaving the olfactory space and reentering it later. So when I have a particularly tough odor to identify, the phenomenon of adaptation enforces a rest: simply sniffing *more* isn't going to bring it to mind.

Practice brings new methods of searching for the odor's name. One extremely familiar smell (many were "familiar") causes me to put down the bottle, close my eyes, and look for the memory attached to it. It is in a house. One of the houses I've lived in. Where? The bathroom? No. Is there another person here? No. I look through a dozen empty rooms in my memory. This, I'd guess, is what it might be like not to have any thoughts — living "in the moment" (the wide, vacuous moment). I put the bottle to my nose again and spend five more minutes in the void. Lysol? Sunscreen?

Magic Marker. Out it pops.

Even the most ordinary, obvious smells sometimes require circuitous routes to deliver themselves, smelled — as they are — out of context. I smell one and try to imagine what kind of place that smell might come out of. It's a spice. Cardamom? (Do I even know what that smells like?) It's a spice my mother would use, but only every once in a while. . . . It's an *occasional* spice. I picture the spice cabinet in my mother's kitchen, the shelves made just the width of the spice bottles, each with a rounded safety bar in front of it to hold in the spice kids on their wild cabinet-opening rides. Then I see it. Mom used it to cook some kind of loaf of meat. There: black, pointy — ah. *Cloves.*

Despite the grin-making pleasures of discovery and recognition, over the visits I begin having a physical reaction in anticipation of my arrival at Rockefeller. Just a very few of the hair-raising smells have left their mark. As I settle in with a new tray, my whole body braces for each approach of a bottle to my nose. My stomach curdles; my throat constricts. It is a major leap of faith to bring a small, condensed odor right up to my nostrils: to deliberately raise the bottle to my nose when I know the odor Pandora that emerges, every once in a while, out of it.

After a time, Peggy Hempstead emails me to schedule another visit. I put her off for weeks on end, until my mind begins to forget the close, cloying smells that I might find in her bottles.

On the day I arrive back on Rockefeller's campus, months after my last visit, it is spring perfection. Magnolias celebrate their two weeks of bloom, cherry trees are aflame with blossoms, the air is mild, and the sky is blankly blue. It is a sharp contrast from the last time I left this campus, with winter lingering and the odors of hundreds of smells ricocheting in my nose.

I find Leslie Vosshall on the third floor of a stark glass building, the Collaborative Research Center, wedged between two other, more austere, older buildings. The building hums with the activity of students preparing tables with displays for a "Science Saturday" event to commence in an hour. Everyone is in tie-dyed shirts, less for fashion's sake than to display "a reaction where we show chemical bonding, basically," one of them says of the end product of fiber + dye. Booths on soil bacteria, the microbiome of the subway system, the freezing point of ice cream, and the DNA of strawberries are spaced across each floor. "That's what actually gives dirt its smell, is the microbes," a young researcher effuses to the visitors to his table peering at the petri dish he is presenting them to sniff.

I enter a glass-walled room in which Vosshall, also tie-dyed, her hair in low pigtails, is preparing a display of familiar-looking bottles across five tables for the day's demonstrations. The week prior, she was elected to the prestigious and storied National Academy of Sciences, an acknowledgment of the terrific breadth and skill of her work, from mapping the fruit fly *Drosophila melanogaster*'s olfactory circuitry to discovering the odorant gene that makes humans appealing mosquito meals; today she is numbering large index cards to be used by visitors to guess the odors they will smell. An industrial fan is on in the corner, because, her assistant says ominously, scents "tend to gather."

Straightaway, she thanks me for participating in what was apparently Study 0780. "It's brutal," Vosshall confesses. Apparently the other subjects and I faced 480 odors presented at two concentrations, with twenty control odors as a warm-up — partly to ensure that everyone could smell — "because five percent of the people who come in are blind to smells."

The study had been, in essence, basic science: investigating how one goes from a molecule to an experience of odor. In Study 0780, most of the bottles held single molecules, not necessarily known smells. "We were just looking at the chemical structure," Vosshall says, "to say, *What does it smell like?* There is no rational explanation for why things smell the way they do." Light comes in wavelengths, across a spectrum, and we have visual receptors tuned to different components of that wavelength, determining what color we see. For smell, there is no sense of what kind of odor "spectrum" or space exists.

"If we didn't know what the rules for vision would be, we would bring in one hundred people and have them look at cards and tell us, *This feels blue to us. This feels purply,*" and that's what Vosshall and her team were doing with odors. The evaluative words we were given was a way to give us a preliminary vocabulary to talk about smells.

"The words are very imperfect," she admits. Indeed, I had often wished for more descriptive options than the seventeen they included.* Some odors were cloying, savory, or gymnasium-like. There were definitely *soft* smells and sharp smells. Pit-of-the-stomach smells. There were pointy smells and fresh smells, and I began to feel as though many odors indeed had a color. There were moments when, in all seriousness, I jotted down, *That is sweet acid tree.* Only later did that description seem like the product of a well-addled nose.

I ask her about those very, very bad smells that stuck in my nose for days. "What's probably happening there is one of two things," she replies, looking down at the cards she is numbering *4.* "The odorant — the particular receptor or receptors it's binding to — causes persistent neural activity. Basically it's bound to the receptor and it won't let go. And the receptor is still saying *gross gross gross gross gross.* The other possibility is that, there are all sorts of proteins dissolved in your mucus in your

*edible / bakery / sweet / fruit / sour / acid / chemical / sweaty / musty / urinous or ammonia / decayed / wood / grass / flower / spices / fishy / garlic

nose, whose job it is to clear odorants — make the *5*s so they're clear *5*s," she interrupts herself to instruct an assistant. "They're overwhelmed, because there's too much."

Avery Gilbert calls this a noseworm — the olfactory equivalent of an insipid tune that gets stuck in your head on constant replay. He speculates that it could also be a kind of "precognitive sensory rehearsal" of the smell — as one might "rehearse," or ruminate on, new flavors or sounds, especially odd or unpleasant smells. In which case, had I continued for a few thousand more bottles, maybe I would have gotten over it.

Before I leave, I stop to smell the first display they have prepared for the day's visitors: vitamin jar–sized containers with cotton balls stuffed inside, saturated with classic smells. Synthetic vanilla — the first synthesized odor, in the nineteenth century — is terrifically recognizable: I name it on the first nostril. I find *cis*-3-Hexen-1-ol awfully sharp, definitely grassy; the accompanying label identifies it as an insect attractant given off by cut grass. I make the discovery that I am in the 30 or so percent of the population who does not strongly smell androstadienone (from which androstenone is synthesized), a component of men's sweat and a potent mating pheromone in pigs. The nearly identical R- and S-carvone molecules turn out to smell similarly to me, not distinctively spearminty or caraway-like. I head outside, looking for a newsstand at which to buy some spearmints, to try to imagine them with a caraway overtone.

There seemed no way I was going to smell as well as my dogs do. And I could lay the blame broadly. My bipedalism might've shot my chances. Also the very anatomy and physiology of my snoot. Worst of all, I had paid no attention to smells in my life, and they were in need of attention. But whether it served to improve my nose or just to get me sticking my nose close to things, the result of this self-experimenting was that I began spontaneously sniffing the world.

I stick my nose in wineglasses before swilling; idly waiting in line at the grocery store, I nose various aisle-lining packages; when the season turns I open drawers to new clothes and take in their hibernating scents. I begin keeping a journal of daily smells. Early entries are pithy: *July 24: peppermint. July 31: horse dung. August 10: sudsed sidewalk.* They expand with time: *March 21:*

sugar house — can smell it before step in, wood smoke and steam of sugared water evaporating; sweet, dense, hazy. On airplanes, I notice, smells are contained and bounce over to me more frequently: almonds next to me, coconut lip balm, airplane food heated up, my hands of the bathroom soap.

One morning I smell a fire on leaving my building. We google *fire, Upper West Side* and get nothing. Later we learn there was a building collapse at 116th Street and Park Avenue, two miles away, minutes earlier. Plausibly, the human sense of smell was originally useful as predictive of toxins and danger. Could I be tuning in to that primitive state?

Many detections are unintentional and, indeed, unwanted. I walk into my classroom before teaching a morning seminar and immediately know that someone has eaten garlic the night before. When I tell them this, they all blanch: even if it wasn't them, could I know something about them by walking by and catching a drift of air? I decide not to report how obvious cigarette smokers are when they enter the room, long after putting out their smoke. I smell a friend's bad tooth before he tells me about it.

In summer, I begin entering our budding garden with nose up, searching for something fresh, resinous, honeylike, sultry, or melon; for a hint of spicy, piney, spring, coarse. My nose catches Band-Aid, tomato stem, and turpentine, with a base of fetidness: inside of a closed cottage, mud, still pond.

After a hiatus to clear my nose of any echoes and to still my stomach, I would begin to concertedly attend to smells again. First, though, I determined to watch the experts.

7:
Nose to Grindstone

I'm crouched under a desk in a back room in an old DuPont chemical building. The building is giant: three stories of impossibly long hallways and once-bustling laboratories. It is now abandoned. The room I am in had been an office, perhaps, for one of the researchers who worked in the adjoining lab. Stripped wires hang from the ceiling of the lab; chemical hoods and faucets collect filmy dust. Everything about the room is decrepit: stained carpet underfoot; stained and broken ceiling tiles overhead; stained and broken Levolor blinds half-mast over casement windows on the walls, unopenable from the inside. The entire building is basement-cold; the halls bare. Though lights flicker on here and there, every room down the long hallway is unkempt, as if the building had not only been cleaned out, it had been cleaned out in a hurry. Bookends are scattered near shelves; drawers are left half opened and half empty. I look on the floor around me. A scattering of small washers peers up at me, yelling *oooooo* silently.

When the door closed behind me, it did not feel good. The room is an insulated chamber. I feel a bit as though I will need to be rescued. I hear a trainer's muffled yell — *"Dog on site!"* — downstairs, and then I wait, hands gloved and clinging on to a leather tug toy. *"Gus on site!"* the trainer shouts. Gus is a yellow Labrador retriever with a taut body and velvet ears. I sit in silence, three stories and a long hallway from where he would have entered the building. Then, suddenly, I hear him charging through the halls: I can almost see the muscular body of a dog through his galloping vibrations in the floors. I think of suspects being pursued by a police dog; I get a glimmer of nervousness. When the dog closes in, when that strong, determined animal approaches, I experience a little of what it might be like to be the destined prey of a skilled predator.

He thunders into the next room. I can hear his breathy panting and sniffing. My predator-cum-savior barks twice, then stops. Then two more barks. Rummaging sounds come from his side of the door. I place myself fully by the door, ready to open it, tug toy in hand. More barks — three, four. At eight barks I open the door and we encounter each other. At the sight of me his eyes seem to widen; he shifts suddenly back. I thrust the toy into Gus's mouth, as I've been directed to do. He takes it, half play-bows, half barks, and tugs with force and pleasure. He's found me.

Working dogs are the real smelling experts. While there are working dogs who do not use their noses for a living — therapy dogs, Seeing Eye dogs — many are "detection" dogs. What they are detecting is some particular thing that has a smell: a lost person, an illegal drug, an invasive plant, a cancerous cell, a destructive insect. Vision and hearing might aid the search, but olfaction is the driver.

But no dog begins life with interest in sniffing out palm weevils. Dogs begin life as wiggly puddles of puppies, eyes and ear canals closed, unable to walk in a co-ordinated way, mostly seeking out the warmth and soft touch that represents both mother and food. To discover how a wiggly puddle of puppy becomes a highly motivated, serious detection dog, I had to follow their noses from the very beginning. I went to the Working Dog Center.

In late summer, the walk from the train station to the University of Pennsylvania Veterinary School's two-year-old Working Dog Center is a journey in shedding comforts. From the cool, spacious vault of Philadelphia's 30th Street Station one walks through redly-bricked Drexel and Penn, bejeweled in leafy trees, before veering south through a more industrial, impersonal medical building corridor and finally out onto a broad auto viaduct where pedestrians are scarce. A 1929 draw-bridge, apparently left untouched since 1929, looks slightly ajar, and offers a view of the Schuylkill River and train tracks beneath it. I am its only crosser. Across the river, an unpromising intersection appears: two gas stations flank a highway entrance. But this, I've been told by the Center's founder and heart, Dr. Cindy Otto, is the penultimate landmark: one is nearly arrived.

By the time a well-warmed and -moistened visitor is waved through the gates of 3401 Grays Ferry Avenue, the accoutrements of organized, polished campus life have been thoroughly left behind. For here stand the remains of the twenty-three-acre DuPont research and paint-testing plant, which closed in 2009 and was bought by Penn in 2010, but has barely changed since the chemists split the scene. The footprint of the site was originally the Harrison Brothers Chemical Company, makers of sulfuric acid and paint. Today, broad parking lots are punctuated by lonely trees giving little shade. Warehouses loom over empty footpaths. A three-story industrial building, long and broad, smells of another decade. The large chemical buildings seem to have spawned a smattering of single-story buildings, windowless or without windows anyone would look into or out of.

Around a corner there is movement. A yellow Labrador puppy is prancing along a dusty stretch of landscaping. She mouths at her leash, at a wafting leaf, at the hand attached to the leash that pulls the leaf from her mouth. This puppy is a sign. Now peer through the weeds and past the squat abandoned building: you might see a black Lab charging across a parking lot, landing and stopping on a cooler, then pivoting, leaping down, and charging to another. Over there, behind that wire fence, espy a broad spread of detritus and abandoned objects — concrete bits and broken wood pallets, a totaled car with its trunk ajar. But look closely: there is a German shepherd dog there, racing up a hill of rubble and disappearing into its bowels. The abandoned site teems with dogs.

First Day: "Seek!"

The Working Dog Center (WDC) is run out of an entirely unprepossessing low building shoved between parking lots full of dormant transportation vans. An identifying sign can be found — by squinting. Before finding any sign, the glimpses of dogs and the din of barking orient a visitor to the door.

Inside there are double gates — to prevent unwitting escapees — and audio cacophony. Kennel cough has hit the Center, and the afflicted, which is now the greatest majority of the dogs and all of the puppies, are kenneled in the front room, their barks resonating well against bare floors and walls. Cindy Otto, professor of veterinary medicine at Penn, appears at once to greet me. This is her center, her brainchild: a training center for dogs who will become detection dogs and a rehab center for dogs already on the job. Otto seems undisturbed by the noise. After all, these are not disobedient dogs; they are just about the most obedient dogs one could meet. As I will soon see, these are super-dogs, even in their youth — already able, or becoming able, to find a person hiding in a three-story building, to locate someone trapped in a broad pipe under rubble, to pick out malignant plasma from benign.

Otto wears a shirt with the Center's motto: "The art and science of working dogs." While it has engaged in classic scientific research programs since it opened its doors in 2012, the Center's staff has always included a number of highly accomplished trainers and dog handlers whose skill comes from their practice, not from reading up on original research. "We want to capture it," Otto says of their knowledge. "Steal it. Reproduce it." To take the practical wisdom of the handlers and reify it, make it policy. "Eventually will that t-shirt say 'The science and science of working dogs'?" I ask. "The artful science," Otto corrects me and smiles a broad, confident smile.

Wait a minute. Let's pause. There are, today, thousands of working dogs all over the country. Dogs working in airports, at the borders, in police departments. Bomb-sniffing dogs, drug dogs. There are ever-increasing numbers of dogs working in conservation settings or in pest control. There are dogs who are able to find cancer cells at rates as good as or better than medical methods. Isn't there already a fully proven science to creating a detection dog? Isn't there already a standard method to determine who is ready to work and who is not? While there are federal and state certification standards, today, Otto says, "There is inadequate scientific research to validate optimal training or testing methodology for detection dogs."

There is no standard procedure for training a detection dog. Nor any one way of testing whether they're good at it. What Otto's Center is doing is working to change that. Much of the work here is preparatory, encouraging certain behaviors and fitness to increase each dog's chance of success once he is assigned a job.

The social center of the facility is a small kitchen with doors on either end. Someone has left a birthday cake (for a dog) on the counter; trainers walk through slowly and coo in admiration. A double-wide sink holds well-chewed Kong dog toys and various bowls and coffee cups. Bags of marshmallows sit next to bags of dog food on a nearby table; a jar of peanut butter sports a handwritten label: PEOPLE PEANUT BUTTER. PLEASE DO NOT USE PLASTIC BAGS AROUND PACKER, a pinned announcement reads, in urgent capital letters. HE EATS THEM.

Around the corner a hallway serves as a makeshift operations room; a whiteboard on the wall lists the dogs' names and various projects and notes — *Hydration; Puppy run; Foster visit.* Pat Kaynaroglu mans the board; trainers are in movement, in and out, gathering leashes and fanny packs to fill with handfuls of kibble. "I'll take Osa, Rookie — the medium dogs," Kaynaroglu says, moving through the day's assignments. No one acknowledges her out loud, but everyone has heard her. In three minutes the hallway is empty and each person is en route to her dog.

Kaynaroglu seems to be in perpetual motion. Each day she is arrived before I come and remains after I leave. As the Center's training manager, she takes a decidedly hands-on approach: she is rarely without dog in hand, hard hat on head, or walkie-talkie crackling. She wears her shoulder-length perfectly-blond hair pulled back in a ponytail, sunglasses propped atop her head. She is a search-and-rescue handler, but also, since almost the inception of the WDC, she has been among a few vital heartbeats of the place.

I follow her outside to where the puppies are playing. It is little but a fenced, dirt enclosure with a plastic pool under a tarp. No matter. Five yellow Labs, eleven weeks old, are delightedly gallivanting. Various volunteers and trainers circle them, grinning at the puppies' antics and pulling things out of their mouths without skipping a beat. These puppies are the WDC's first litter. While breeding working dogs has long standing — and many facilities will get their dogs only from specified breeders or breed lines — the advantage of raising their own puppies is that the dogs can be observed, molded, and shaped from the very beginning. Dogs destined to be working dogs are not trained before they are twelve or eighteen months old. So while they are growing up, it seems reasonable to try to prepare them for their future life, as well as to fend off any potential problems or direct them toward the kind of job they may be best suited for.

In front of us are the comically named Pak, Packer, Patterson, Parsons, and Pinto — the P litter — at this point discriminable only by the color of their collars. Even as we watch, though, we begin to see small differences in behavior: Parsons (pink collar, female) is often the leader of the troop; Packer (dark blue collar, male) is trying to put his mouth around *everything,* dogs included. In addition to these five, Punches, Philip, Pierce, and Pearl are living with foster families. They will visit the Center once a week, but, by living apart, serve as a kind of control group to this uncontrolled experiment of raising the puppies. The Center group will have early training and assessment and daily social rumbles. The fosters will be living, well, as pets. Whether that will make them less suitable to future employment remains to be seen.

For now, the puppies are playing. Later in the week they will have their first assessment of their developing personalities. Kaynaroglu's walkie-talkie squawks and she heads to a broad lawn where a dog has splayed himself on the ground, his tongue hanging long out of his mouth. Two men take measurements of him; he does not protest. Kaynaroglu approaches: "How're you doing, buddy?" She smiles at him.

Ohlin, a shiny-eyed chocolate Lab with a glossy coat, is taking part in a study. In essence, handlers play fetch with the dogs until they are worn out; then their temperature and blood work are taken, to see the efficacy of different kinds of hydration. "One hundred forty-four," a technician reports his heart rate when Kaynaroglu appears. A well-slobbered ball lies next to Ohlin. He has stopped retrieving, a straightforward behavioral marker of *hot.* Kaynaroglu takes a look at him and says,

"One-oh-five?" The technician smiles: "one hundred four point nine." She has guessed his internal temperature. "I look at their respiration, their tongue — is it cupped, how far out is it? — are their eyes squinty?" she tells me. "Easy things — how long they have been working . . ." The outside of the dog tells her about the inside.

Another trainer walks by, noting a graffiti of urine on the sidewalk. "Was Jesse here?" she asks. She could recognize the dog even in his absence.

All of the staff have this kind of second sense about the dogs. They know where they've been, sense their temperature, and talk about dogs' dimensions as though they are part of each other. And, indeed, for the time that they are at the Center, the dogs' lives are supervised and overseen by these acute lookers. Everyone is always watching the dogs' behavior, assessing them, and the chatter is all about their latest suspicions and working hypotheses about the personality, abilities, and inner life of each dog.

We head over to the hulking old DuPont laboratory building. They call it "227," for its former address. It is here that I will hide, to be rescued or apprehended by Gus. At its entrance, Annemarie DeAngelo and Bob Dougherty, two experienced handlers, hand out assignments: a person to take videos; another to keep track of the various tug toys; a hider, the "target." DeAngelo, the WDC training director, is retired from the New Jersey Police Department; Dougherty is a volunteer here, still an officer at a local police department. They radio over to the Center to have someone bring the first dog by. As the designated hider, Dougherty slips inside the building before one of the "pointy-ears," as everyone refers to the young Dutch and German shepherd dogs, is brought over. Felony, seven months old, is small for her breed and has a dark, lovely face. A handler holds her close. When Dougherty's in place, he radios over the all-set; Felony's handler says *"Seek!"* and seek goes Felony. Though new to this game, she has already connected her handler's exclamation-point enthusiasm with getting to run through the building, and so she does, handler in tow.

On the first floor there is a draft down the hallway out the open door. Felony pulls strongly and pokes her nose in a few rooms, but then heads to the stairs. Everyone trundles up after her. She is straining at her leash. The stairwell is greatly musty; the air is close. Smell has been deeply carved into the walls. On the second floor she pauses, and her handler encourages her to look in one of the first rooms. Dougherty pops up from behind a door and gives her his arm, with a bite guard on it. At this delicious offer, Felony lunges.

After watching a few dogs seek — and being sought-after myself — it dawns on me that there is no "smell training," as such, happening at 227. Though no dog is born knowing how to search for a person, what the trainers are doing is not teaching them to *smell a person out.* The training assumes that the dog *already knows* how to do that. That comes built in.

Instead, what needs to get developed is the drive to search for that smell at any time, at a person's behest. To search when there are dogs nearby wanting to play; to search past piles of toothsome smells and cooing people; to search when one seems not to be getting any closer to the target. To search simply so that at the end one gets to tug on a leather toy or the moving arm of a target.

Search training begins with small steps, bringing the dogs to the building's front door and having the "target" appear from behind it, then naturally evolving through small increases in duration to looking in rooms, encouraged by the handler holding them on leash. The dogs catch on quickly. Over the course of an hour I watch one pup go from playing with an octopus-shaped tug toy outside, to following it down the hall, to finding the toy (held by a trainer) behind a closed door. At each step the dog is rewarded for his "discovery," however large or small, by getting to play with the tug or wrap his mouth around a slobbered tennis ball. Later the leashes will be dropped, the searches made steadily more difficult. The dogs will be asked to "alert" — to tell their handlers that they have found the hidden person, usually through barking.

Barking is not natural for all dogs. The trainers keep calling Rookie, another pointy-ears, a possible "dad" dog, which I take to mean one that will make some father a good family pet. Later I learn this is shorthand for diabetic alert dog (DAD) candidates: dogs with good noses, but not as much drive to pursue a hidden scent like gangbusters. And in Rookie's case, no particular inclination to bark with excitement when she does find its source. We watch as she finds hidden Dougherty (upstairs closet, behind three doors and around a corner) with ease, but then just sets her rump on the floor and waits. She has a classic three-month-old-puppy shape: large head and ears, small gangly body with a hint of muscling developing underneath. Patiently, she stares at the door that conceals him. Eventually, she looks back at us, the observers half hidden outside the room, quietly hoping for her. Her expressive eyebrows mark concern, endearingly. Perfect pet dog behavior. But Rookie is not being trained to be a perfect pet dog. We wait. She waits. Finally, she lets out a single, shrill note — a request, a plea. A bark. Everyone cheers, and Dougherty appears, sticks a tug toy in her mouth, and lets her play.

This is the unstated mantra of the design of the working dog: create an anti-pet. We want our pets to be motivated to sit still, not tear up the house looking for a scent; to look back at us in request, not bark endlessly; to keep their mouths closed in enthusiasm, not to bite. At the Center, the trainers build up those bad habits one by one, assiduously rewarding them. Some dogs need training to bark to alert. The pointy-ears are working on "bite" — which is just as it sounds: bite and hold the person (for now, the covered arm of a person) they find. Each of those natural behaviors — search, bark, bite — is highly useful in the context of work. Each had a function for the dog in the species' evolution — functions that are often moot for pet dogs. In training dogs to work, we are giving them reason to use those behaviors they evolved.

Inside the Center you can see the result of this anti-pet development. The rooms ring with successfully trained barkers. And while in most contexts people are the focus of the dog's gaze, here the dog seems to be looking beyond the person to something . . . some smell . . . else. The dogs are not vying to be pet or tickled or to curl up next to you on the floor. Nor does anyone but the occasional visitor even try to pet or tickle them.

Neither are they, after puppyhood, especially interested in dogs. On their regular walks around the outside of the Center to pee, they are hustled along, encouraged to ignore other dogs' scents, distracted by obedience games and the fanny pack full of kibble that brings their attention to the handler. Dogs are walked separately, run on searches separately, and kenneled separately and often with visual barriers to even seeing dog passersby. As we wait in front of 227 for the next dog, a good-natured yellow Lab and his handler approach — the handler attuned to his every pause and suggestion of loitering: "We're not gonna mark here; we're not marking." She suddenly sees us and, like an owner with a difficult dog, turns on her heel and walks away.

While the dogs certainly know of each other's presence — the sound of dogs is ubiquitous; the smell of dogs hangs in the air — they are learning that other dogs are not the most important thing in their lives. As they must not be later, in their working lives. To the uninitiated, the Center looks like it is running an elaborate game of dog tag, with dogs being moved around but rarely catching up to each other.

Instead of encouraging smelling or sociality, these dogs are encouraged in their *drive.* Drive to pursue the scent, the toy that arrives at the finding of the scent — and to be frustrated when they cannot get at it. In many ways the dogs are being kept (when working) in constant vacillation between immense frustration and satiation. There is something they hugely want — usually a game with a toy — and they have to wait and wait and wait until they can go after the scent that lets them get it. Many times I have seen the dogs shaking with excitement to get started.

SECOND DAY: "FIND DOPE!"

I follow Bob Dougherty into another, smaller DuPont building. Dougherty is an active K9 handler and comes to the Center on his days off. He wears serious work shoes and an easy manner. His comfort around dogs is apparent.

PApa Bear, a full, strong chocolate Lab, is by his side.* Today he will be initiated into the world of narcotics: PApa Bear is being trained to sniff out drugs. He has been a generalist thus far, and Dougherty is tasked with trying to get him to be more of a specialist, so that he can be usefully placed with a permanent handler. Up a set of worn stairs, through a dizzying set of doors and small anterooms, we arrive in a big double-sized room with a chem-lab kitchenette along one side: heavy on the sinks. The room has a huge multicolored sunburst on the floor, some late-night "paint test" accident.

Dougherty takes out a large plastic bag and pulls out a towel. It has been cavorting overnight with a smaller plastic bag containing a small amount of the scent source. Quantity matters: a dog trained on a subtle

* All of the Center's dogs are named in homage to someone (dog or person) who was on the scene at the World Trade Center explosion and its long aftermath. PApa Bear is named after a working dog named Bear from Pennsylvania ("PA").

scent may not know to alert if he has discovered a large amount of the scent: it's almost a different stimulus. On the other hand, starting with a large amount of the scent in the room is too much. "If I put two pounds of marijuana in here," Dougherty explains, "the odor will be all over the place." A dog just learning won't be able to localize it.

Even so, I can easily smell the presence of groggy, giggling teenagers in the park as the bag is opened. We can all smell it, surely. The task being presented to PApa Bear, though, is to learn that this sweet smell — more or less, stronger or weaker — is the smell he needs to find in order to get a partner for a vigorous game of tug.

Without being able to explain this to PApa Bear, Dougherty simply works backward from the desired result: he *begins* with a vigorous game of tug with the towel, rolled on itself and secured with rubber bands. As PApa Bear mouths the towel intensely, his strong tail slicing the air, Dougherty tells me, "Dogs can eat this without getting ill. Not so with other drugs — we use pseudo-narcotics for training there. Plus, drugs like heroin are sometimes cut with rat poison" — nothing you'd want your dog keen to put in his mouth. Then Dougherty tosses the towel; PApa Bear readily pursues it, bringing it back to him for another tug. After this, Dougherty begins to feign-toss the towel, making PApa Bear work to find the source. The dog is thrilled, breathing heavily with excitement and effort. But when he begins to search he closes his mouth until it is barely ajar, his head cocked in contemplation. Already, something has happened: the dog is "going to get" the odor and "bringing it" to his handler. The start of narcotics search.

Like all the handlers at the Center, Dougherty is continually improvising new variations and new steps, based on knowing the dog, how he's performing relative to other dogs in training, and what the conditions are like — is the dog hot, overworked, distracted, ill? PApa Bear looks to be getting the game, and Dougherty starts ramping up the difficulty level. Within minutes he hides the towel in plain sight, then out of sight, then hides the bag alone. *"Find dope!"* he instructs, introducing new vocabulary to match the new game. PApa Bear finds the target at each step without a hiccup. He's now finding dope. It's been about forty-five minutes since he walked up the building stairs to this room, never having smelled the stuff before.

Once the target is set, Dougherty guides PApa Bear around the room, working on "detailing" — helping him learn that after he's barreled toward an area that has the odor, he should explore the source in detail. For dogs, who are not furniture users (except when they sneak up on your sofa), human-world objects that have surfaces and undersides, insides and partitions are not obvious. By being guided to the top, middle, and low reaches of a counter, for instance, the dog can learn that a general "counter odor" may be different at different altitudes, or inside or behind it. Then when he hits the source — *BAM!* — it's a big, different smell.

We head back to 227 to "try PApa Bear out" in a new space. Dougherty will hide the baggie in a half dozen new spaces before calling it a day and letting the dog rest. I stand in the middle of the first room — one of the old laboratories — with a video camera, hoping both to keep up with the dog and stay out of his way. A pipe with a faucet fitting and a sleeve reading nitrogen hangs from the ceiling, entangled with insulated and cut wires. PApa Bear enters the room running. He cruises all the way to the back wall, nose up, then turns heel and starts careering back. He takes barely a peek at me. The way he examines the counters and each drawer beneath feels habitual, as though he's just checking where he left that file folder or test tube. Only his assessment is rapid-fire. He rears on his hind legs and appears to look at a cloud of air about six feet off the ground. His nose guides him to the northerly wall. Then he pivots, finds a cabinet, goes behind, then returns to the front and barks to alert, staring at the top drawer, his jaws violently snapping shut with each bark as if grabbing at the scent. And yes, PApa Bear, the baggie is inside the top, closed, drawer.

As the team moves on to their next search, I go back into the room to pick up the sample left behind: when it was found, Dougherty tossed PApa Bear the tug and the sample was immediately forgotten. Since I know where he alerted, I stop and sniff the spot myself. The drawer is closed, and I smell nothing. I pry open the drawer a few inches; it is bound by tape — prepped for moving day and then never moved. I still can't smell the sample; nor do I see it. I move to visual search, looking deeper into this drawer, opening the other three drawers to check them. Then I return to the first and look again. The baggie is right at the front; I simply missed it. Drawer ajar, with my eyes on it, in the place where I knew it was, I missed it.

As PApa Bear keeps searching, I wander back past the Center to a field with multiple apparatuses for balance, strength, and skill work scattered around. It smells of recently cut grass. Under a high sun, Kaynaroglu and DeAngelo are standing on either side of a ladder. It is propped at a low angle and leads nowhere. Kaynaroglu holds a ball near its end, though, and Gus wants that ball. To get it, he is on the rungs, tentatively inching forward. He does not look happy to be there.

Teaching the dogs "body awareness" is one of Otto's real missions. Not just to have proprioception, but to be fit, agile, physically confident. Any working dog out in the world is going to encounter uneven surfaces and tricky terrain, to need to negotiate an uncomfortable path. Otto wants them to be ready. Her interest began as a veterinarian, with seeing working dogs with substantial injuries from being unfit or unable to use their bodies strongly.

Over my few days at the Center, I saw the ladder brought out many times. It seemed to stand as the epitome of dog fitness: while a dog may never need to climb a ladder in his working life (although many will), when he can, it is only because he has mastered using his body skillfully.

Or, as the trainers say, "Dogs have to understand they have back legs." Gus might be able to get to the ball because he is muscular and can brute-force his way up the ladder. But they will not let him move forward unless he moves his feet one at a time, settling firmly on a rung and knowing that he has before he deigns to move the next foot. "He has to make a connection between here" — Kaynaroglu points to Gus's rear foot — "and between the ears." A trainer uses a clicker to reward him each time he moves his rear foot at all. I leave the group after a grueling several-minute exercise in which Gus makes it about two rungs. Before the month is out, he'll be walking that ladder like a champ.

THIRD DAY: "PUPPY PUP PUP PUP PUP PUP!"

Today the dogs are in the "Field," a westerly part of this old campus that Center staff say, sighing, used to be much better when it had great piles of dirt on it. Now it has been leveled and is one and a half football fields' worth of cracked concrete, with weeds considering taking it over, and various large containers and broken vehicles scattered about. Large PVC barrels are rolled hither and thither into the Field, blue beacons in the desolation. On the other side of a fence and some marginal wilderness is a sanitation transfer center: one can hear, then see a small bulldozer scuttling along its perimeter. Kaynaroglu looks for the wind direction: there is little in the Field that sways in a breeze, but at once we are hit with the smell of the trash pile. The wind is from the southwest. The smell starts out sweet and becomes nauseating: there is no "adaptation" relief for our human noses, because the wind keeps shifting and bringing new wafts.

The dogs are brought out one at a time from their crates. None seems to notice any of this sensory scene, except the barrels. For today the barrels, or any large item, may hold a person — Kaynaroglu or another trainer, volunteer, or visitor willing to stuff themselves in a hot, close space, pull a lid over the opening, and wait to be searched for and found. Some of the dogs have been here before; for others, this is their first go. But all know that the game will be to discover the person — any person — who is not in sight.

Sirius is out first. He is one of the older

dogs here, at two years; his panting face gives him a smiling expression. *"Go find!"* He races past observers, uninterested in the *already-found,* as he is released into the Field. Kaynaroglu has been hiding in a barrel on the other side of the Field since before he came out. He finds her in forty-two seconds and barks. She tosses a ball on a rope in his mouth and tugs it as a reward. "You're just a model guy," she gushes.

Kaynaroglu moves to a new hiding place while the next dog is brought out. Gus. He starts quickly, tours the grounds a bit more, and suddenly settles down to poop. While doing so, he spots a scent and beelines toward her. "Hi! Nice to see you! Yay!" Kaynaroglu beams. One minute thirty-two seconds. Slowing down seemed to allow him to see.

Next, Jake bolts out, a yellow Lab like Sirius, with asymmetrical ears and a deeply dark nose against his blond fur. As he passes we can see his eyebrows working and hear him whining with excitement. He touches his nose to each barrel, then lifts his nose into the air and licks it. Fishing for molecules on the wind, surely. He runs circles around Kaynaroglu's barrel, barking, thirty-four seconds after his release. "Someone else!" she says. He pivots and finds a second trainer squirreled away in a large pipe, tugging rope in her hand.

The dogs keep coming out, each with his own style, and each a model of efficiency. Quest (bright-eyed, tall-eared young German shepherd): twenty-seven seconds, in a great unhurried lope.

Logan (also a young shepherd male, with soft eyes and a dark face): one minute forty-nine seconds. His sniffing is impossible to see when he is running, but when he turns abruptly, it is the nose that is leading him. *It* turns, and pokes to the ground, or into the air.

Felony (a dusky dark Dutch shepherd, the baby of the group): thirty-seven seconds.

These dogs, finding their way along the Field, are in essence drawing a picture for those of us standing and watching their paths. They are drawing a picture of how air moves around the Field. When Logan begins, he first heads toward a Dumpster and a pile of detritus, far off from Kaynaroglu's hiding place. But this area is downwind of where all the searchees have been, and a distant scent of them must remain. Finding no one, he pokes his nose into the wind. He moves farther out into the Field, and again his nose is up. Finally, he is downwind of Kaynaroglu, in a triangle of space behind a mound of concrete chunks — where a large pool of Kaynaroglu-scented air would have been collecting. Seconds later, he's nosed her out.

They are drawing a picture of the way dogs process smells. By looking at the barrels first, they are using visual landmarks to aid their olfactory search. By retreating from old, deteriorating smells — representing a person who was in a hiding place but no longer there — they are using quantity and age of odors to tell them something about the present moment. By passing by the observers with nary a look, they are using their budding knowledge of what a search is to ignore the very ripe smells of people right in front of them, who don't need to be searched for.

They are drawing a picture of the mind of a dog. Being olfactory is living in an impermanent space, where seeming "objects" — fixed for us, with our visual approach — are fixed only as long as their smell remains.

The handlers walk Quest around the corner, staying on the outside edge of the campus. Unseen, behind a fence, the Schuylkill River wends its way through Philadelphia. On its opposite bank, two of the Center's previous dogs — Socks and ZZisa — now work with the Penn campus police.

Behind a locked fence is a treasure. The rubble pile: mounds of rebar and concrete and slabs; wooden pallets; PVC pipes; miscellaneous junk; a smashed and gutted car, its trunk creepily poised open, as though someone has just casually retrieved his belongings after a disastrous accident. A few corrugated rubber pipes are jammed downward into the mess, like portals to the innards of the earth. The rubble pile is not an unsightly mistake; it is carefully constructed to provide a place for the dogs to practice searching while also becoming fearless and physically flexible. There are dozens of places where a person might be trapped or hidden: the pile is riddled with awkward gaps and crevices, mimicking the disorderliness of a sudden building collapse. If you assent when the trainer asks you, *Do you want to hide?,* you leave the comfort of human company and breezy airflow and may scoot into the slippery inside of a broad pipe. It is too wide and smooth inside to get traction and you settle in, deeper than you'd like, pulling a makeshift lid over the opening. It fits awkwardly, a scratched window showing a piece of azure sky. The bottom of the lid is suddenly the only sign that you are not *truly* lost in a pile of catastrophe-site rubble. From this sepulchral vantage, you will grow sweaty and hear your own heart beating in your ears. Then this sound is replaced by the galloping vibrations of a dog close by. You might see the beautiful moist nose of a dog like Quest appear in the slit between lid and barrel, almost bending to reach in and confirm inside — then *EFF-fuh-fuh-fuh,* quick strong sniffs — then hear an alert bark, twenty alert barks. Good dog.

Over where Gus climbed the ladder yesterday, six puppies tumble over each other, mauling ears, grabbing tails, bumping bodies, and squirming as they wait in a small fenced corral on the grass. Five are the twelve-week-old yellow Labradors; the sixth is one of the pointy-ears, Dargo, a week their senior, whose bearing is less "cuddly puppy" than "detection preprofessional." They are all about to meet their first obstacle course. Suddenly one of the corral's walls has an opening and the nearest puppy — Parsons (pink collar) — hops through it. She has begun the "puppy run," as the trainers call it: an improvised, narrow runway along which she will confront obstacles that require hopping over (a tire on its side) or walking carefully through (a ladder lying flat) or bravely negotiating (a wobbly narrow board, suspended a few inches from the ground) or barreling into (a dark tunnel). The other puppies continue their antics, until one notices the opening, too, and follows Parsons's tail through the hole. Then another puppy, then they all charge through.

A half dozen volunteers and trainers watch the scene. Annemarie DeAngelo stands a little apart. She is another one of the Center's heartbeats. A former law enforcement officer, she developed the New Jersey State Police's canine program. Even out of uniform — "I'm a civilian now," she says — she has a no-nonsense manner. She does not suffer fools gladly. But when one of the golden-haired puppies looks momentarily disoriented, DeAngelo approaches the fence and melts into a high-pitched *"Puppy pup pup pup pup pup! Good girl, good girl!"* to encourage her along.

The puppies are not being explicitly trained at this age. But most of their day's activities are cleverly designed to sneakily prepare them to be *more trainable* in whatever working position they wind up. The obstacle course is designed to expose them to new environments — shadows, uneven or wobbly surfaces, gates — in a safe, non-pressured way, and at an early enough age that they learn not to be fearful of new or unknown things.

Running the lot of puppies through the course also takes advantage of learning that happens between dogs. When one dog charges into a dark tunnel, any puppy watching is more likely to follow him: this is called social facilitation by cognition researchers. Here it is called *puppy see, puppy do.*

"C'mon, munchkins," one of the trainers calls out to the puppies considering the business end of the tunnel. Though the dogs vary in their willingness to move through the course, by their fourth round, they are all barreling onto the wobbliness, over the obstacles, into the tunnel. Simply getting the chance to be exposed to these new things, and watch others do it, has turned a riotous play session into a context for real development.

As tongues hang low, a few of the puppies are taken aside and brought to a quieter area, a sidewalk leading to 227 where a trainer is waiting for them, clipboard in one hand and a basket of Very Strange Objects in the other. This is the puppies' first assessment test. They are not being assessed for their skill at smelling. While the volunteer who whelped the puppies and housed them for their first weeks did expose them early in their lives to odors — a cut-up tennis ball, leather, other novel scents — to see how they reacted, it is not yet clear that subtle differences in smelling as a puppy lead to differences as an adult. In the meantime, these trainers are using an early test of "drive and courage" designed and used by the U.S. Border Patrol. In its first stage, a puppy is simply presented with various loud or surprising objects, and the trainer makes a qualitative impression — from one to five — of his reaction.

One of the litter wags merrily as the trainer shakes a little bottle with coins inside, tosses a rattle under an upturned Tupperware container, drops small metal pipes on the ground, drags keys along a metal ramp. An umbrella is suddenly opened and dropped on the ground. The puppy races after the pipes, paws the Tupperware, follows the keys up the ramp until she catches and mouths them, and is unfazed by the umbrella. For now, fives all around.

FOURTH DAY: "ROTTING MEAT AND PROM CORSAGES"

As I walk to the Center on the last day of my visit, the wind carries an odor of sulfur along the river. This will not be the worst smell I smell today.

Today we go back inside 227. "Do you want to run Gus on HRD?" Kaynaroglu asks Dougherty, who already has the leash and is heading out. I jog after him.

One day, this part of campus will be renovated into housing or new facilities for the university, but for now, this old industrial building is a great training surface for the Center's dogs. Not only is it expansive, with hundreds of rooms inside of rooms; the rooms all present different olfactory challenges. "You work here awhile, you start learning about all sorts of air currents" in the building, DeAngelo says. While she can't articulate what she knows about the air currents, all the long-time trainers have learned that certain rooms and times are more difficult for the dogs. The way the air flows determines whether one can get a sniff at the doorway of someone inside, or whether their smell loiters and collects in a corner of the room, discoverable only by close investigation.

One room in particular — a lab on the second floor, old chemical hoods in its center and closed doors on the perimeter — is especially difficult for the dogs. Over an hour, I watch three dogs pass the door concealing Dougherty multiple times before clearly paying a different kind of attention to it: what might be called *examining*, instead of browsing. When they do, concertedly sniffing, they all find their man straightaway.

"We sometimes spray baby powder to follow airflow," Dougherty says: in essence, using a powder gun to see what the dog sees. In other words, to get a sense of how air moves in the room — where the smells might be lingering, where they are rushed upward or straight out a window — one can follow a dog, or one can put something into the air. In the Field, the dogs were our source. Here, a powder gun or the like makes visible the invisible (to us).

If you reflect on it, we have all watched airflow. The mesmer of a lit match comes from its phosphorous flame, to be sure, but also from the curl of smoke that slips up and away from it. We are seeing airflow when we watch the "fog" used in theater ooze across the stage and increase the moodiness of a scene. Smoke and other particulates, though, dissipate when air is turbulent — which is much of the time. Other methods use a "neutrally buoyant" agent, like helium bubbles, which slide into the slipstream of air. With a specialized light, this method can be used to trace air as it moves naturally around a room. In an ordinary empty room, air warmed by the day creeps up the walls onto the ceiling, where some is bumped downward again and crashes, wavelike, into air coming from the other side. The two streams tumble together, paints swirling on the canvas of a hallucinating van Gogh.

In essence, that is the path of odors, too: the path that these dogs see in every room they walk into.

Dougherty holds a small glass jar. We are on the second floor of 227, hiding the "HRD." This turns out to be one of those initialisms that is lovelier in brief form than long form: human remains detection. In the jar are a few pieces of human remains: bits of knee from a cadaver. The supply catalog for these folks must be fantastic.

"Do you want to smell it?" he asks, unscrewing the top. I do not, but I do. I lean over the jar, instinctively pulling my hair back from my face. At first I detect nothing. Then: a sweet and vomitous waft, a carnation rotted in a bin. Dougherty smiles at me. This is one of the only times one knows just what is in someone else's head.

Upton is in the distance, over-interested in something on the ground. I hurry over. His nose points at a recently deceased squirrel being visited by a small number of flies. His tail wags and he weaves his head around, mimicking rolling in it. I ogle, draw him away, but then go back. I must smell it. I bend over. It is not difficult to smell. Sweet and foul together — not the kind of smell that is appropriate for a squirrel who recently was chattering and racing up trees.

Most people might not be attracted to it as dogs are, but there is a fascination and a depth to the smell of death, of things rotten and gone. Hemingway described the odor's many parts as though it were a perfume: combining the kiss of a woman who has tasted blood; the morning streets mixing last night's cigarette pails and wastewater with the soap used to clean them; and decayed flowers; all experienced with the swimmy belly of a seasick ship passenger. This "saccharine putrescence" — P. J. O'Rourke writes of *eau de corpse* as invoking "rotting meat and prom corsages" — sickens us even more for our knowing the cause: death sickens us.

Mortuary technician Carla Valentine says,

"You *have* to smell what's going on in an autopsy because you can diagnose things with smell . . . no professional examiner of the dead would ever try and *mask* the smell." For a mortician, the "natural smell" of decomposition is not so much offensive as it is appropriate for the context. Context is, if not everything, highly determinant with smells, at least in one direction. Any enjoyed odor smelled in the wrong context — lavender in our coffee cup; smells of the barbecue on our beloved — can become foul.

What the dogs are smelling, whether they are intent on a decomposing animal in the park or are HRD dogs, is death: death of cells, and the odor of the biological processes that take over as cells are broken down. That dogs find it notable makes them particularly good death-detectors — and may also enable some of their abilities to notice diseased cells in their living owners.

Dogs' apparent interest in the smell of death does not mean that they will instinctively lead you to the dead smell in the room, though. At the Center, HRD is introduced just in the way all the new scents are introduced: in baby steps, slowly shaping the dog's behavior first to notice the scent, then to pick it out from other scents, and then to bother to tell the person about it. And at the end of every search, a tennis ball or a tug toy awaits. The dog's reliable satisfaction with his payment is written in the strong wag of his tail.

A year later, I write to Otto to check on the dogs. She reports that PApa Bear, who had his first exposure to marijuana in a paint-splotched building with me, is now a drug and search dog for the Gloucester County Sheriff's Department in New Jersey. Gus, who found me and found his rear legs, is working as a certified search-and-rescue dog in New Mexico, as are Felony and Sirius; Jake is doing search and rescue in Pennsylvania. Quest and Logan are police dogs with SEPTA, Philadelphia's transit system; Rookie is also a police dog. The former puppies Packer and Parsons are search-and-rescue dogs; Pinto is an HRD dog. Pak and Patterson are trained in search and rescue and awaiting the right team to bring them on. A new litter of six — the Q litter — is now at the WDC learning the ropes.

Spending the better part of a week with these dogs infects you. When I return home to my lovely, goofy, non-ladder-walking dogs, who sniff my eyes closely and look to me to solve problems of tennis balls under couches, I also see how much they are not workers. This fact, to be sure, cuts both ways: though unemployed, our dogs are engaged in heavy bouts of mutual adoration and silliness with our family. On the other hand, this seems to be their only occupation. Is it enough?

What impresses me most about the Center's dogs is how much they are treated on their own merits, dealt with as members of a species with specific skills — and not treated as furry, quadrupedal humans. A dissatisfaction I feel with my own field — comparative cognition, broadly conceived — is that it begins with the premise that the most interesting topic of investigation is whether nonhuman animals can perform at the level of humans at various tasks.

By contrast, the WDC dogs prompt me to turn the comparison on its head. Given what these dogs can do — find a stranger hidden in a three-story building in two minutes; learn the smell of a target odor of any sort in a single session — what I think is more apt is seeing if *we* can do what dogs do.

And so I aimed to follow what some of these detection dogs are doing further — and then ask if people, too (including me) can learn to detect what the dog does. Comparative cognition, written by the dogs.

8:
Nose-Wise

As I rise slowly up out of sleep I become aware of Finnegan, his nose millimeters from my mouth, snuffling. I open my eyes and he looks surprised to find me attached to whatever-that-smell-is. I mentally note to call the doctor.

-Dog-

In the UK, in the late eighties, a Border collie–Doberman mix noticed something. Her owner's left thigh. At the same time, her owner, a forty-four-year-old woman, began noticing her dog's extreme attention to a mole that had cropped up — on her left thigh. Minutes would pass while the dog examined the mole with her nose, even through her pants. When the weather warmed and she dressed in shorts, the dog began nipping at it, as though to bite it off.

A few years later, a Labrador named Parker began paying particular attention to *his*

owner's left thigh. The sixty-six-year-old man was constantly pushing the dog's nose out of his pants leg as Parker tried to get at the patch of itchy eczema that had developed there.

In the United States, a Dachshund puppy began showing unusual interest in his owner's left armpit. The forty-four-year-old woman, in good health, endured some time of this sniffing while they shared the couch in front of the TV. One day she moved the dog away and, feeling the area, discovered a lump.

In each of these cases, the object of the dogs' attention was a malignant cancer. The first case resulted in the discovery of a melanoma — which removal may have saved the woman's life. The man's eczema turned out to be carcinoma. And the axillary lump was biopsied and discovered to be breast cancer. After a mastectomy, the woman still had the Dachshund's attention at her armpit. The woman had radiation and chemotherapy, but, a year later, died of her cancer.

As the authors of a paper reporting the first case wrote in the medical journal *The Lancet,* "Perhaps malignant tumours such as melanoma, with their aberrant protein synthesis, emit unique odours" — a kind of signature scent — "which, though undetectable to man, are easily detected by dogs." Before this case in 1989, the idea of dogs detecting — essentially, *diagnosing* — cancer would have been laughable. But these dogs, all of them untrained to do anything remotely medical, were able to contribute to their owners' lives — in some cases, lengthening their owners' lives — simply through being themselves.

There is little better proof that the nose is how detection dogs are able to detect their quarry than when they are after that most invisible and sneaky of villains, cancer cells. Now, the authors of each of these reports were careful to describe their case studies as "only anecdotal." But the stories were provocative. Moreover, the standard business of cancer detection was expensive, lengthy, and sometimes painful. The hope of replacing office visits, biopsies, and CAT scans with a dog scan, so to speak, was irresistible. Those first anecdotes birthed a small industry of research studies designed to suss out exactly what the dogs were noticing, and if they could be trained to notice it on anyone at all. A cancerous growth certainly produces compounds that are volatile, and that would be effused in blood, urine, or breath. Except for the strangeness of dogs wagging into the medical arena, there was no reason that dogs wouldn't be able to become diagnosticians.

At the Penn WDC, another trainer, Jonathan Ball, is "doing cancer." In other words, he is running dogs in trials designed to entrain them on the smell of cancerous cells, and to train them to alert to that smell.

The door to the training room opens and Ball lets me in. He has a boyish haircut and is dressed in jeans with a treats bag at his hip. The room is dominated by a specially designed training "wheel" — set on its side, with twelve radiating spokes. At the end of each arm sits a small glass vial, the size of a diner saltshaker. They are covered by mesh screens to conceal any possible visual cues — and to prevent overenthusiastic lickers. In three, there are minuscule donated specimen samples. The target vial has fifty microliters of plasma generated from a pooled group of cancer patients with malignant ovarian cancer. Two contrasting vials hold the same volume of either benign-tumor patient plasma or normal plasma. The remaining nine jars are empty.

I peer into the vials. They all appear empty: fifty microliters is not a visible quantity. I smell the pizza from the kitchen next door; I smell the astringent tang of isopropyl alcohol, with which they clean the wheel; I smell the soap on my own hands; I do not smell the plasma. A long, lean yellow Lab, Ffoster, is gracefully padding her way through the kitchen with her handler. En route from her crate through the kitchen, she noses the food bins extra carefully. She takes a quick tour of the room: the single cabinet, the two people sitting discreetly along a wall. Ball turns to her, and her attention is on him at once. He plies her with treats, saying *"Focus!"* until she is looking at his face. She actually seems to look right into his eyes, maybe three inches into his skull, her pools of brown eyes turning inward somewhat, reflectively. She ever-so-slightly tilts her head. *"Go find!"* he says, and she heads obediently for the wheel.

Ffoster sweeps her jowls smoothly across each mesh top, taking less than a second. Without breaking her trot, she repeats her analysis with the next one, until she pauses, just barely, at one. She looks at Ball, who is staring unwaveringly at the due center of the wheel. His eyes are narrowed, his posture fixed. He is "blind," as they say, to the position of the malignant sample in the wheel — only another person in the room, back turned to him, knows where it is —

but he is also being careful not to cue the dog in any way. Since the time of Clever Hans, the horse who was alleged to be doing arithmetic but was actually just a highly skilled reader of his trainer's inadvertent body language, anyone who trains or studies animals has been preternaturally attentive to whether they are "cuing" the animal. Experiments ensure they are always blind: that they do not know which cup the dog is "supposed" to tip over to get the hidden treat, or which person the clever dog "should" beg to get a reward. Similarly, in training, one wants the dog to learn that the cue the trainer devises — a word or gesture, often — is instruction, not some accidental sound or movement. By contrast, many pet dogs make themselves seem very smart indeed precisely *by* noticing owners' inadvertent cues. Hence their magical knowledge of when it is time to eat, walk, sleep, go to the vet, have a brush, or have a bath: they are reading their owners, and their owners are happy to be read.

Ffoster gets nothing from Ball and continues a very light pace around the wheel. She traces it twice, carefully, but does not alert. Then she sneezes and sniffs extra hard, almost speaking *Hmm-huh-huhn-mm-huhn,* then *Phrunh* and sits. Alert. The knowledgeable trainer assents, and Ball breaks out of his statue stance and rewards her for her find.

Ball sends Ffoster out of the room to have a quick lunch. "We have to remember not to do cancer at lunchtime," he says. He considers her slight delay in response to be evidence of her distraction by the dogs' doled-out lunch helpings and four boxes of pizza she passed in the kitchen en route over.

"C'mon, girlie," her handler says as she is led back. Time for another tour of the wheel. Ffoster's training is an exercise in *shaping:* gradually encouraging a desired behavior by rewarding all the subcomponents leading up to that behavior. To teach a dog to water-ski, you don't set the dog down on the beach to watch other water-skiing dogs, or drag the dog out into the water with you unawares. Instead, you begin with the smallest of steps: getting the dog on the ski — on the beach. Simply approaching the ski might earn a treat. Once the dog is approaching reliably, you delay the reward until he has stepped on the board. When he is happy stepping, ask him (by delaying the reward again until he acts) to step with two feet, or four. Soon enough a dog can be taught to stand and stay standing on the ski, without ever being explicitly asked, or, indeed, even knowing what he is doing. In later steps the "water" part will be slipped in. When a dog is water-skiing, the behavior has been "shaped."

Here, too, Ffoster and the other dogs are led through days of shaping. Instead of water-skiing, they are only being asked to care about, and eventually find and sit next to, one particular smell. Ball and the other WDC trainers begin by presenting the dogs with a sample of the diseased plasma (combined from many patients). If the dog sniffs, he is rewarded with a click and a little treat. Sniff, click. Sniff, click. Eventually the diseased plasma will be brought out next to plasma from healthy persons. In that case, the game is to sniff the diseased sample, but not the healthy sample. Sniff, reward. If he sniffs the other sample, no reward. Repeat. They do this hundreds of times, but each session is short so as not to exhaust, discourage, or bore the dogs. Once they get to finding the sample at the wheel, training continues: ten times at the wheel, twice a day.

Eventually, dogs like Ffoster and McBaine, a black and white springer spaniel with the yearning look characteristic of the breed, become so proficient that when they miss a sample, the trainers know it's not because of the dogs; it's because of something else: a growling stomach, a budding cold.

After Ffoster leaves, McBaine comes in with his handler, Annemarie DeAngelo. Between dogs, the room is dust-mopped, the bins and mesh spray-cleaned with isopropyl alcohol. "Where are we?" DeAngelo chatters with McBaine as they enter. "What are we doing?" McBaine sniffs the non-wheel parts of the room: the floor leading from the kitchen door to an outside door, the chairs for visitors, a visitor (perfunctorily). "Hi, dog," she says, to get his attention. "Are you going to sniff around? Hi, goober."

After he settles, DeAngelo steps behind a screen — Clever DeAngelo — and says *"Seek!"* McBaine sets to it. He addresses the wheel with a trotting, not overly rushed pace. He nearly rests his muzzle on each bin, making full swipes across the mesh, leaving moist streaks on its top. Long eyelashes frame his face; his feathery ears droop. One can't help wondering if he is hoping to find a partridge in one of these vials. On the first trial he sits, alerting, to one of the "distractors" — the normal plasma. No reward. The vials are emptied, reloaded, and he goes again. On the second trial, same thing. Finally, on the third he hits his stride, easily finding the sample, and earning a delighted response — and a tiny edible treat — from DeAngelo as reward.

Later, via Skype, I watch Tsunami — *Tsu,* as everyone refers to this handsome, long-eared German shepherd — run through her training. She is "reactive," the trainers say — meaning, in this case, she isn't fond of unfamiliar people or dogs around, so I watch from an adjacent room. Even on the small computer screen Tsu is formidable. She cruises around the wheel like a circus horse circling the ring, seeming barely interested. Her tongue dangles lazily. Then suddenly she turns, as though to address the wheel, and finds the sample. DeAngelo tosses her a ball on a rope and she proudly prances with it. Then she circles the wheel nine more times, finding the sample each time with barely a hiccup or an eyebrow-raise.

In the WDC protocol, the dog needs to be at 83 percent specificity to go to the next level — that is, correctly ignoring the noncancerous samples ten out of twelve attempts. This level of performance is statistically highly distinguished from random guessing, which would have them correct one in twelve times. Indeed, in the next months the dogs are finding the correct sample (sensitivity) and not alerting to the incorrect samples (specificity) more than 85 percent of the time. But . . . why are they not *perfect*?

A puzzle. The first plasma they encounter is "pooled," combined from many patients, so it could be that training on pooled samples makes it difficult to generalize to an individual sample. It could be that the dogs are picking up on some, but not all, of the smell profile of the samples.

Or it could simply be that the game is not sufficiently interesting to them *every time.* Humans rarely achieve 100 percent success at even the skills and tasks we have long since mastered — hence the ubiquity of tripping while walking, hearing your own voice saying *runny babbit* or other such spoonerisms, or having to look up which way daylight savings time goes every year. We get distracted, have bad days, and grow sleepy; a working dog does, too. The dogs are being only human here.

Many other dogs, at many other research groups, have been tested on their cancer detection skills. Though the training has been less assiduous than at Penn's Center, the results are stunning.

■ ■ ■ ■

Research studies on the topic have, for the most part, shown dogs to be eminently good cancer detectors, via a range of means and substrates. Given that there is not one cancer but many cancers — in lungs and other internal organs, on skin, in blood — researchers generally take samples related to the source.

To few people's surprise, dogs were plenty interested in the cups with small amounts of urine that they were presented to sniff. Perhaps to more people's surprise, six mixed-breed dogs were able to distinguish bladder-cancer patient urine from healthy urine at rates better than chance. Urine, carrier of the final throes of whatever metabolic processes have been happening in the body, seems to have a different cast when there is disease in the body as well. Prostate cancer, too, might leave traces in urine; in one study, a Belgian Malinois used it to correctly spot the disease 91 percent of the time. Two German shepherds retired from explosives detection spent six months instead sniffing urine cups from hundreds of patients and healthy volunteers; at the term's end, they were nearly perfect in identifying the cancer patients' samples.

The studies go on, testing dogs on tissue samples from biopsies and even right on the person: another two dogs were terrifically reliable at finding the melanomas from among thirty bandages placed on human volunteers. Perhaps the best medium for this research, though, has been exhaled air. Linus Pauling, who won both a Nobel Prize in Chemistry and the Nobel Peace Prize, was less well known for, in 1971, discovering that there were hundreds of volatile organic compounds in an ordinary person's exhaled breath. By those who study it, "bad breath" — or even "breath" — is not a simple phenomenon. The constituents of breath represent a couple of things: what is in air that's being breathed in, and a gaseous impression of the metabolic process that has happened in the body before being exhaled. It now appears that everyone's breath is different, reflections of themselves and their innards. Some two dozen compounds are identical across people, and then another two hundred or so are unique to you, from among thousands of compounds researchers have caught in their sticky tape.

Breath also holds information about disease in the breath-organ, the lungs. To collect breath samples, researchers ask people to exhale a few times into a test tube in which a bit of polypropylene "wool" has been stuffed. Just as smoke sticks on clothing and chlorine in hair, the volatile compounds in breath stick on the wool. A top is snapped on the tube and it is sealed in a zippered baggie. A few weeks of clicker-training, and one study's five young dogs — Labrador retrievers and Portuguese water dogs being trained as guide dogs for the blind — were snappily finding the samples from patients with lung cancer.

Again, though, a puzzle: what is it, exactly, that the dogs are smelling? Diseased tissue might have hundreds of volatile molecules; is there one that is the cancer "tell"? If so, it could be isolated and presented, without the bother of actual cell samples. And can we be sure that what dogs are detecting is the disease at all? A body fighting disease will have inflammatory and immunological responses — which themselves might emit smellable stuff. Accompanying illnesses — and even the depression or anxiety that might come with a serious disease — might also be the noticed scent.

At the Monell Chemical Senses Center, George Preti is trying to find the silver bullet. Having moved on from his research on canid anal sacs, he is now analyzing the arguably lovelier plasma samples that the WDC dogs are training on. When I visit him, on a hot afternoon in August, West Philadelphia, where Monell is located, smells dazzlingly pungent. Vents on the street fume with the fury of smelly, imprisoned dragons in the sewers below. Monell is blissfully antiseptic by comparison. I speak to Preti in a cool room lined with bookshelves with titles on flavor, taste, and olfaction: *Sugar Research 1943–1972* jostles for room with *I Was a Food Writer for the C.I.A.: A Dietary Confession, Hunger: A Biopsychological Analysis,* and *Umami: Proceedings of the Second International Symposium on Umami.*

As I puzzle over how he is analyzing the component parts of the smell of cancer, Preti perks up. "Well, I'll show you." He pushes his chair away and starts out of the library into the hallway. I follow him downstairs into a classic-looking laboratory room: beakers, tabletop lights, piles of partially boxed equipment scattered about. He stops at what looks like office copier equipment that you hope you never have to learn how to operate. I would need to make no copies. Instead, I was about to meet the most thrilling technology you've never heard of:

"This," Preti says with moment, "is a gas chromatograph."

A large, sealed safelike box with an elaborate keypad on one side, its front piece opens, microwave-like, to reveal its innards. "It's not on now," he reassures me, opening a thickly insulated door. What it does is, in the science of smells, nothing short of futuristic. Insert a sample of something smelly — an orange, an old book, a violet, the captured airspace above a baby's head — and the "GC" provides a listing of all the volatile molecules found in it.

"The guts of it is really the column," he says — a column that holds fifty meters of coiled glass, half a millimeter in diameter, and is coated along its walls with a polymer. "So. You inject the sample up here" — pointing to the top of the GC — "into the hot injector" — kept at between 200 and 300 degrees. Helium gas is pushed through the coil, and the sample materials and the polymer interact. "Then you gradually increase the temperature of the column at a set rate, say four degrees a minute. As you go up in temperature you start to volatilize a lot of the things that are condensed on the column. They all start to proceed down the column — but they proceed at different rates" according to their different molecular weights. The components are thus separated, a chord pulled apart into an "arpeggio," as Avery Gilbert writes of the process.

"Now this particular GC is set up as gas chromatography (with) olfactometry." Preti points to the output nozzle. As the solution is separated into different compounds, each compound emerges at a different time. "If you have a complex mixture, and you want to know when certain odorants are coming out, or certain kinds of smells, you can sit here and sniff and record your olfactory impressions," he explains.

I am a little startled. This fancy machine, helium, polymers, and temperature gradients, and at the end there is just someone sitting there . . . *sniffing*? "Oh my goodness. You must be trained to do that . . ." I suggest.

"Well, I could train you to do it in five minutes. It's easy. Just sit here and sniff. The problem is applying the right vocabulary."

He shows me a piece of paper by the nozzle with a list of possible smell descriptors, three columns long, including quantity characterizations (*slight* or *nothing*); familiar foods (*pizza, pickles, hot dog, popcorn, coffee*); and qualitative impressions (*swimming pool, clean*). "Olfactometry" is "person sitting with nose in nose-shaped inhaler at nozzle's end, smelling."

Alternately, "GC" ends in "MS": mass spectrometry, in which the different compounds are identified by a machine that graphically represents the compounds as they emerge. The smell is translated into an image; the compounds, a series of drawn peaks on paper. The bigger the peak, the more of that compound is in the mix. The GC-MS, then, tells you all the volatile — potentially smelly — components of any substance: coffee, lilac, citrus, soil. If you put urine in the GC, it will come out as many different compounds as well, including those of the foods eaten by the urinator.

What the GC-MS does not tell you, though, is which compound is the smelliest — to us or to dogs. The biggest peak is not necessarily the biggest smell; nor do all the peaks contribute equally to the smell you experience. In some cases, a tiny trace component is actually responsible for the odor. Our sense of *coffee* might be mostly captured by a mere handful of the six hundred or so compounds separated by the GC.

As of the day we spoke, Preti had not yet found the compound that makes the cancer patients' plasma samples distinctive to the dogs. Should he find it, that biomarker would not only make training dogs more efficient; it could transform human diagnosis of disease as well.

The inside of the dog's snout, with receptor cells grabbing passing volatile odors from the sniffed airstream at different places in the nose, is in essence acting as a biological gas chromatograph. The olfactometer at the end, analyzing the samples, is the dog's brain. Oh, to get a glimpse of its list of smell descriptors.

Given dogs' abilities, does it make sense to put your book down, strip, lie on the floor, and ask your dog to sniff you for melanoma tout de suite? I cannot recommend it. He may indeed detect that errant mole on your torso — but he does not know to tell you about it. It may smell different to him, if he is familiar with your smell, as any dog who spends time on the laps, couches, or beds of their persons is bound to be. It may even smell "diseased": one dog-olfaction researcher proposed to me, tongue not entirely in cheek, that wolves' sensitivity to the weakest or most ill of their prey may be related to disease-detection dogs' reactivity to human diseases. But the dog does not have a concept of disease; we must draw that for him. If he sniffs a skin ailment, he will simply notice it and say nothing.

It is not only cancer that has captured the attention of detection-dog trainers. Work is well under way training dogs in detecting hypoglycemic and hyperglycemic episodes (in diabetics) as well as in alerting prior to an owner's epileptic seizures. One research paper describes dogs, glowingly, as a "fully biocompatible and patient friendly alarm system" for hypoglycemia.

At Penn, the trainer I saw working with the cancer-detection dogs, Jonathan Ball, designed a special "diabetic suit" — coveralls with pockets sewn variously over them into which samples of the odor could be stashed (and searched for). "What are they smelling?" I ask him. "Good question," he replies, shrugging. Nonetheless, at the time of my visit, the Center had just, to some fanfare, placed its first fully trained diabetic-alert dog, Bretagne, a sweet-tempered young golden retriever, with her new owner. Bretagne's training began by imprinting her on the smell of saliva swabs from someone whose blood sugar level was in the 50 to 70 ml range — on the low side of normal. For nondiabetics, this blood sugar range is experienced as the slow, sluggish feeling that might descend about half an hour after a big meal, when the insulin released for a moment overtakes the glucose level it is regulating. In training, Bretagne is shaped to alert at a blood sugar level of about 80 ml, as it is falling. The test swab and control swabs are put in vials and pocketed, and Bretagne ticklingly sniffs the coverall-wearer all over until she reliably spots the one.

But *what it is* in the saliva sample — what volatile odor, exactly what compound or set of compounds that has the indicative odor — is not yet known. Just as with cancer, until the signature cocktails of disease are known, biological samples will need to be used.

In Philadelphia, once Bretagne went to live with her person, the training process only shifted. Weekly training is needed in detection so that the dog continues to get some reward for alerting. Unlike other working dogs, a diabetic-alert dog may have no "successes" — positive identifications of precariously dropping blood sugar — during the day if her person is stable. Although that's good for the owner, it can be frustrating for a dog: trained to do something, and do it well, she wants to get a chance to succeed. Training sessions with samples off the body keep the dog's motivation high and nose focused.

In the UK, the first empirical test of the success of diabetic-alert dogs was published in 2013. The first indication that dogs might be clueing in to their owners' health came from *owners'* reports of their pet dogs spontaneously responding to their hypoglycemic episodes; stories with so-called epileptic-alert dogs have similar evidence. Now, owners may remember only those instances in which it seemed the dog warned them — not those when the dog did not, or when he showed these behaviors without an episode under way: a foible particularly typical of all humans. If that were the case, the levels of owner-reported dog success may have been overreported. So the researchers asked owners to take blood glucose samples of themselves throughout the day and also to note when they believed their dog alerted — by nuzzling, pawing, even retrieving the blood-test kit. It turned out that the times when these dogs alerted was more likely to be out of the desirable blood sugar range than the other routine sample times. Dogs did not get all the hits, nor did they always hit correctly, but it was promising. Dogs might have been smelling a change in their owners' sweat or even breath — but also, as acute observers of their people, might have been noticing some change in behavior that served as corroboration.

In the cases where these detection dogs are living with their focus of detection attention — their people — as opposed to other working dogs, it could very well be that some of the life improvement of diabetics and epileptics is due to the general positive effect on well-being that living with a dog can impart. A person who had never been able to be left alone could now be alone (with her dog); people restricted from travel would now be free to travel. Dogs did that.

-Person-

Dogs' capacities at cancer detection are, to my thinking, the perfect example of what seems most mysterious about the dog nose. Cancer is not only something we fear or dread; it is something opaque to the mind and often invisible to the eye — certainly not smellable. To the dog, freed of our conceptions of mortality and our obsession with sight, cancer is just a smell.

On the other hand, a bit of our own, mostly neglected history is relevant here. It is this: humans have been diagnosing disease by smell for thousands of years. Only in recent times have we essentially stopped smelling.

So here I must go back in time. We have grown so antiseptic, so machine-addled, that we no longer bother to sniff (or, sometimes, look at) patients. This has not always been so. Ancient thinkers and cultures were aware of the role of smell in disease. Smelling one's patient and his diseases goes back at least to Hippocrates, who advised doctors to keep an "open nose." The ancient Greeks thought about smell as symptomatic: to Plato, smell arose in transformation between two elements: "As water changes to air, and air to water, all odours have arisen in between." The physician Galen characterized smells on people, from their mouths, as either "according to nature" or "against nature."

The interest in smells led to some bizarre ancient medical treatments. The belief that a pregnant woman's uterus could travel up to her throat and suffocate her led physicians to wave horrible-smelling material near the woman's mouth (to send the womb away), and pleasant-smelling stuff around her genitals (to draw it near). Pliny noted that the "rank" smell of the armpit was thought to be addressable by drinking a white Falernian wine — and then urinating out the smell. Whether this method replaced the smell with a smell of wine, or merely made the drinker sufficiently besotted that he did not notice his smell, is unknown. Other recipes involved chewing gum; perfumes, too, were thought to be health-giving.

Linnaeus's original eighteenth-century classification of the fundamental kinds of scents was actually medically inspired: he was interested in botanical smells and the curative (or ill) effect they would impart on their consumer. While a "fragrant" plant (lime, lily) might be healthful, "repulsive" or "nauseating"-smelling plants — such as veratrum, highly toxic — were clearly to be avoided. Plants with no odor at all were no good.

Despite Hippocrates' "open nose," the fear that bad odors themselves caused disease loomed large. Galen warned people to avoid "those who exhale such putrid humours that the houses in which they lie in bed become stinky." Given Hippocrates' continued influence in modern medicine, perhaps the contemporary worry that some still hold that foul odors breed contagious illness is not surprising. One helpful sixteenth-century instructional manual for doctors suggested readying oneself when approaching an ill-smelling patient with a fragrant branch of burning juniper and speaking to the patient "from a certain distance away." If one must approach, it suggested, venture forth with back turned, and take the patient's pulse by groping behind you with your hand, never getting any closer than necessary — and with an assistant there to hold the branch directly underneath your nose.

Nonetheless, smelling diagnostically continued. After some research on guinea pigs, breath-sniffing had become a medical tool (little is said about the breath of the guinea pigs). In the eighteenth century there even developed a medical specialty — called osphresiology, forgettably — that reified a list of characteristic smells of disease. Odors from the skin as well as bodily effluvia — vomit, urine, sweat, fecal matter — were all informative. A sweet smell of onions indicated smallpox (something that distant cheetahs are apparently drawn to, should you see a cheetah sniffing after you)*; freshly baked bread, typhoid; the smell of a butcher's shop was a sign of yellow fever. A

* "It is stated that the odour emitted from a smallpox patient will attract cheetahs from afar" (*The Lancet,* 1906): while this may seem far-out, we are certainly accustomed to the *flies* attracted by our smell.

fruity, ripe-banana or "fake fruit" smell indicates diabetic ketoacidosis. Dozens of conditions and diseases have henceforth been added to the list. Smell of cat? Ringworm. Tickle of freshly plucked goose feathers in the air? Measles. Stale beer, sour bread, old straw, and sweet, urinous, or putrid smells are all indicators. Even psychiatric disorders may have characteristic smells: pungent and sweaty, urinous, and vinegary odors have been smelled in schizophrenic, psychotic, and anxious patients, respectively.

Often, odors can indicate the ingestion of something toxic: arsenic smells like garlic on the breath, iodine is metallic, mothballs or eucalyptus reek of camphor, and any glue-sniffing is exuded from the body as the smell of petroleum. Cyanide can smell like the bitter almonds it can come from; alcohol poisoning, the alcohol imbibed: juniper berry (gin), fermented grape (wine), hops (beer).

By the nineteenth century olfaction was not unusual in medical diagnoses. Infections, in particular — sepsis, tooth or bone decay — were common, and carried distinctive smells. The tide then turned. Professional interest in smells rapidly diminished as infections — and their pungent odors —

were reined in. The decrease in olfactory diagnostic techniques coincided with the steady increase in the stigma of smells and smelling. Many cultures grew concerned with ridding the environment of smell, and, coincident with that, technologies that could do the job of "smelling" bodily effluvia began to replace human noses. In the twentieth century, the gas chromatograph supplanted all but the keenest of noses, with its ability to separate and identify the molecules in a complex odor — be it from mouth, armpit, groin, or palms.

Today, little attention is explicitly paid to smell in Western medicine. While doctors and nurses seem to regularly refer to "patient odors" (especially malodors), what those odors may denote — apart from some unpleasantness — is getting forgotten.

When I ask physicians and nurse practitioners how they use odors in their practices, I get remarkably similar regrets: "unfortunately . . ."; "alas"; "sad to say . . ."; "sorry." Dr. Jane Orient, editor of the masterly *Sapira's Art and Science of Bedside Diagnosis,* which has an entire section on the "bouquet" of odors one might detect from a patient's breath, tells me that she doesn't know of *anyone* who uses smell prominently in their medical practice. "It is a much neglected subject," she says.

Contemporary medical texts mirror this waning of interest in smell as diagnostic. One exception is in the realm of infectious disease. Infection is, simply, the state of being affected by some non-endogenous organism, like a foreign bacteria in an otherwise healthy body. Infectious-disease textbooks write about smell as though it were a natural sense to use in diagnosis. Indeed, when trying to figure out if an anaerobic infection is involved, "Only the foul or putrid odor of a lesion or its discharge is specific," one states; other clues are only suggestive at best.

"The characteristic foul odor of the sputum suggests anaerobic involvement" is a not-unusual sentence in medical literature. "Almost all doctors have come to know that a really foul smell" — of cadaverine or putrescine, say — "coming from the lung or an abscess or a wound represents the presence of anaerobic bacteria," Dr. Bennett Lorber tells me. "Foul-smelling" is one of the major categories of medical smelling, apparently. Even the petri dishes of bacteria that pass through microbiologists' hands — Lorber is a professor of microbiology and immunology at Temple University's School of Medicine — might be smelled. Some evoke grapes; some bleach; and some have a "mousy smell," Lorber adds. While specimens are now usually examined by machine more than by nose, if Lorber gets ahold of one of those agar plates himself, he sniffs.

So doctors *are* smelling, to be sure. The same practitioners who claimed to not wittingly use smell in their practice all casually mentioned some ways they in fact notice patient smells. "Alcohol, of course; some poisons like organophosphates; uremia; hepatic failure; blood in stool if a patient has a big GI bleed . . ." The list tumbles out of Orient. Others mention the distinctive smell of the combination of alcohol and tobacco, which might indicate that other drugs are being used. Dr. Abraham Verghese, who has written and spoken widely about the value of bedside-exam techniques also demurs when asked about the use of smell in his practice. The next second, though, he observes that he did notice odor "in the course of (medical) rounds, incidentally — (such) as the fruity odor of diabetic coma."

What there currently is not in Western medicine is a practice of *teaching* smelling for diagnosis. It would not be impracticable: Orient recommends that a battery of vials with substances like chloroform, the carbon tetrachloride in fire extinguishers, and other toxins be kept available for sniff-training. In fact, emergency room workers at Bellevue Hospital were once trained on a "ten test tube sniffing bar" holding the characteristic odors of poisons. In the 1970s some medical schools used similar means to teach students about possible odors, but its use seemed to wane quickly. It is hard to imagine why, given the possible smell training field trips. What doctor would decline a visit to a brewery, to become familiar with the oaty smell of oasthouse urine disease, an inherited metabolic disease, or to reacquaint themselves with the smell of hops absent the drinking of it?

Sure, along the way, the observant doctor will lean close to a patient and notice what that intimacy reveals. But Western medical schools largely ignore smells. There are machines for that.

Enter Eastern medicine. Or, rather, reenter: the medical traditions from the East far precede Western practices, of course. And in traditional Chinese medicine (known as TCM), smell has always mattered — and still does. The corporeal body itself is the source of information about what is wrong with the body: the means are asking, looking, touching, smelling, and listening. Noting the breath, sweat, saliva, mucus, urine, feces — and even the odor of a room that's had you in it — are part of TCM's diagnostic method.

Turn off the main drag in Northampton, Massachusetts, and the character of the town comes through. Bay-windowed Victorians with small, well-tended lawns; an ambling citizenry; a vegetarian restaurant that does not allow patrons to wear fragrances. On the corner of State and Center, two flights up, is the office of Leta Herman. Her waiting room is cool on a hot day. Incense burns and I hear the trickle of water. Herman practices acupressure and what she calls the Five Element strain of Chinese medicine, which differs from TCM in some of its approach.* But boy, can she smell.

"It took me eight years to get to this," she says happily of her olfactory acuity. Herman has a positively ebullient air: she approaches smile first, her curly brown hair framing a youthful face. "I was not a dog going into it." When Chinese medicine gave her relief from a stubborn medical condi-

* For instance, the language of TCM is more likely to refer to the "Five Phases of change" (*wu-hsing*).

tion, she enrolled in a class to see how it worked. Soon she quit her job in the computer industry and now, sixteen years later, she owns her own practice. And smells a lot.

In TCM's conceptual scheme, five fundamental constituents — wood, fire, earth, metal, and water — are invoked as contributing to the harmonious balance or imbalance of a system (such as a body). To practitioners like Herman, the elemental types *smell*. That is, they are considered to have distinctive odors, similar across people despite differences in race, age, cleanliness, interest in perfumery, or recency of running in a polyester tracksuit. Odor descriptions can be blunt — a person may smell burnt, rotten, or rancid — and sometimes impressionistic: some people effuse the effervescence of "ginger ale or vinegar," an odor that "hit you in the nose — and it's like *pow*, and then it disappears," Herman says. Or the smell of "clothes on the line, drying"; a honeysuckle blossom that "follows you down the street and entangles you."

After talking to Herman for a while, I have to ask: "Do I smell like something?"

Herman offers me a table to lie down on, then softly lifts my wrist to take my pulse. She looks at me penetratingly — smelling at me, I realize. Many of her patients don't know that she is smelling them: "I don't really bring it up with people," she says. Given the strangeness of being sniffed — as well as our own self-consciousness about our body odors — that seems a sensible choice. She and her colleagues have developed tricks for getting a sense of someone's smell without actually burrowing into the person's neck. For one, they leave the room for a good amount of time — twenty minutes should do it. When they return, the person's odor has filled the room, and — *boom!* — it hits them in the face. Another trick for some of what Herman describes as "dense" smells that tend to descend (something like "the smell of vitamins, or gym clothes that have been in the gym bag a long time") — is to "accidentally" toss a pen on the floor under the treatment table. In bending over to pick it up, one can cast about for any gym-bag/vitamin notes in the air.

She furrows her brow slightly, fixing her gaze somewhere in the middle of my forehead. "I'm getting a little hint of water," she says. That means an ammonia smell — *putrid,* in the jargon. I nod, trying not to look too aghast. Herman switches to the other side of the table and gives a big exhale: "You haven't been in here very long." So at least the room isn't filled with the smell of urine. "I could smell your neck if you want to know your [foremost] element. It's kind of cheating."

I agree to cheating. Sitting up, I experience the unusual interaction of having a stranger intentionally lean in toward the back of my neck for a sniff.

"Hmm. Hmm. No, I don't think it's water, then. Yeah. Okay, so metal and earth are predominant."

I ask what use it is knowing someone's smell. "Then you can tell if their smell is *off,*" she says. "So I might be looking at their stomach and notice *Oh, this is getting sour; it's fermented; they have smelly feet.* Their metal smell is getting really rotten; their sweet earth smell is sickly sweet. Then I'm trying to bring them back to their good smells."

Surely TCM is not the only bastion of human olfactory diagnosis, I reason. So, smelling of earth, I dig into the medical literature. A few more reports of smell diagnostics are still bouncing around in the twenty-first-century journals. They share references to some of the now classic smells — cyanide, the fruitiness of diabetes, the "sewer breath" indicating a gastrointestinal infection or periodontal disease. They also share a kind of surprised delight at discovery. These doctors and nurses have become olfactory Sherlock Holmeses, able to sniff out something invisible and elusive in their patients. An alcoholic trying to avoid drink smells fruity, betraying his consumption of the cleaning fluid isopropyl alcohol, a severe central nervous system depressant. A corn farmer who is sweating and vomiting smells of garlic — indicating that he's been affected by the pesticide he was applying to his crops. A couple of coal miners arrive at the hospital unconscious. Their odor of rotten eggs — as well as the blackened coins in their pockets — reveals that they inhaled deadly hydrogen sulfide underground.

Medical personnel are still encouraged to follow their noses if they suspect some of the conditions that produce distinctive odors: phenylketonuria (PKU), with its musty, murine odor; maple syrup urine disease, pancakes-evocative; infections; toxic ingestions. Early suspicions that lead to further investigation can preempt development of serious problems or death.

But remaining evidence for disease smellers is few and far between. As an example of few, George Preti, at Monell, though not trained as a medical doctor, turns out to do his fair share of sniffing people. "You'd be surprised how many people write to me and say, *Oh, my head stinks, it oozes, it comes out in my hair.*" Preti tells them to come in having not washed their hair for several days. Then? "Oh yeah, I smell 'em, yeah," Preti says. "We sniff the patient. It's probably the only place on the planet where people are diagnosed this way." From head odor to concerns over having a very bad body odor, people find their way to Monell. "Oftentimes people are way out of line in terms of how bad they think they smell," he says reassuringly. One of Monell's research interests is a metabolic condition called trimethylaminuria (TMAU), an uncommon and debilitating disease that can easily be identified by smell. Why? It is also known as fish odor syndrome.

And far between: in 2012 an untrained Scottish woman reported that she could smell her husband's Parkinson's disease, a motor disease that often involves sensory changes. He smelled suddenly "musty," she was quoted as saying. When tested on her ability to diagnose which of twelve t-shirts — six from Parkinson's patients, six from healthy volunteers — were from Parkinson's patients, she got an impressive eleven correct, incorrectly identifying one as having the smell when he did not have the diagnosis. The collars of the shirts, where oily secretions from sebaceous glands tend to collect, seemed to be the culprit.

As for that one incorrect t-shirt? Its wearer was diagnosed with Parkinson's eight months later.

This woman's ability forms an interesting twist, because smell *is* being used diagnostically in a different way. In particular, when a patient has a sharp decline in his own sense of smell it is itself indication of possible disease — especially the early stages of Parkinson's. Alzheimer's, too, though a disease of memory, is often marked by olfactory dysfunction. Apart from self-reporting smell loss, patients can take what is now a standardized olfactory test. It uses "scratch-and-sniff," a microtechnology more often associated with foul-smelling stickers and ill-advised novelty books. To make scratch-and-sniff patches, tiny capsules holding slurried oils are sprayed onto paper, waiting to be scratched — and the capsules burst — by an aging fingernail.

As I consider Finnegan's close sniffing investigation of me in the morning, I rummage around in the closet and unearth a forty-three-year-old board game named, punnily, *Smell & Tell,* billed as a "scent-sational game" that is centered around scratching-and-sniffing cards. Incredibly, odors of banana, chocolate, root beer, and garlic have waited, patiently, for decades before meeting my family's apprehensive noses. Finnegan, for his part, will not suffer the game and retreats to a distant room.

(Though Finn had not, happily, sniffed out a serious illness on me, some consider *pregnancy* to be a disease — and he had detected that, I found.)

The German philosopher Friedrich Nietzsche, a man fond of smelling (he once wrote, "My genius resides in my nostrils"), spoke in an offhand way about the "keenness" of one's nostrils: an ability to smell something out, rather than just smell something. This kind of wisdom of the nose used to be reified in the English language as the word *nose-wise.* In the twenty-first century, it has lost its precocious, keen-nosed meaning and is now just defined as "with regard to the nose." Some demotion. Nose-wisdom has a place still in our society; it just needs to be given the respect that the dog's nose is getting.

9:
Stink-waves

The otherwise cheerful musical film *Chitty Chitty Bang Bang,* starring Dick Van Dyke as a mildly dotty Caractacus Potts, takes a dark turn about halfway through. The two Potts children find themselves holed up in a barony where children are outlawed, and whose baron employs a "child catcher" with a spectacularly long and bulbous-tipped nose who can *smell out children.* In no time at all he sniffs the guileless kids out, then lures them with candy into his candy-wagon-slash-cage. In retrospect, it is no surprise to learn that this film script was written by Roald Dahl.

Dahl wrote preposterous noses. Of one of the eponymous, child-hating subjects in *The Witches:* "She can actually smell out a child who is standing on the other side of the street on a pitch-black night." When a child protests that he is reasonably clean (having recently bathed), he learns, "It isn't the *dirt* that the witch is smelling. It is *you.* The smell that drives a witch mad actually comes right out of your own skin. It comes oozing out of your skin in waves, and these waves — stink-waves, the witches call them — go floating through the air and hit the witch right smack in her nostrils."*

-DOG-

Detection dogs (known in various dogly circles as sniffer dogs, detector dogs, or working dogs) detect much more than disease, of course. Just how much more has not yet been discovered. Until recently, we assumed that the dog's nose was only as good as our imagination was in applying the nose to finding things. Our imagination has proven more fertile of late — hence the expansion from training dogs to find drugs and land mines to orienting dogs toward mealybugs, smuggled agricultural products,

* To the great delight, no doubt, of Dahl-reading youngsters everywhere, the tactic to avoid witches is to "never have baths" — thus curbing the stink-wave-dispersal with layers of dirt. Alas, in Dahl's version of the fairy tale *Jack and the Beanstalk,* Jack only escapes being smelled out by the giant ("FEE FI FO FUM . . .") *by* bathing, emerging "smelling like a rose."

and endangered ribbon snakes. Dogs have been trained to find minute amounts of environmental contaminants: derivatives of gasoline production and various toxic products at industrial disposal and waste sites. They work sniffing out sea cucumbers being illegally exported from the Galápagos, and the smuggled ivory and horns unfairly removed from their elephant and rhinoceros hosts.

And dogs detect *us:* tracking, trailing, search-and-rescue, and scent-identification dogs pursue the missing, escaping, lost, or dead; the criminal, confused, unseen, or unlucky. Their noses' authority is acknowledged by our legal system: no less than the U.S. Supreme Court described the detector dog's sniff as "sui generis," a tool unlike any other. Tracking's earliest appearance among canids was no doubt as *hunting:* predators like canids cannot just wait until prey shows up under their claws or leaps into their open mouths. Being able to follow the tracks of (hoped for) prey is a necessary adaptation. Since the time of their wild ancestry, domesticated dogs have modified the hunt into a hunt-minus-consumption. Pliny wrote of huntsmen carrying even their aged and infirm hunting dogs with them, so good were they at finding their quarry by "snuffing with their muzzles at the wind." But in common with those hungry wolves, the dogs find us primarily through one thing: our stink-waves.

The smell of a person is so strong that dogs can follow it over time, underwater, after the person is long gone, and even after the thing the person touched has actually blown up. In one study, researchers found trained bloodhounds able to identify who had touched a pipe bomb — *after* the pipe bomb exploded. The person's scent, laid on the pipe by touching it while setting it up, "survived" when little of the actual pipe did. Dogs have been trained to find drowned persons: the odor of decomposition rises to the surface of a lake or other still body of water; some dogs can even track in flowing stream or river water. Where sonar, divers, and underwater cameras fail, a dog may start pointing at the location from the dock, and then from a boat narrow the search space to just twenty feet across. Cadaver dog handler Cat Warren writes that when a dog alerts during an underwater search, it is "like stepping from one room to another": from fumbling in the dark to a room aglow. Avalanche-rescue dogs have found people buried under twenty-four feet of snow: the person's scent travels to the surface and the dog, after some digging to convince himself, alerts there.

In some cases of search and rescue and trailing, the dog is given an article of clothing to "search out"; but in most tracking, all the dog needs is to know that he is after *some person.* For each of us exudes a perfectly loud and boisterous smell to a dog. What smell-science researchers have shown us is that "without a trace" really is impossible: we are always leaving a trace. A flurry of skin flakes trails from us as we move. Even when quite still, we are wafting out odor from our skin and the things on and in it. Not only that, it is still there long after we are gone. To a dog, wherever you go, there you still are.

Should this sound fanciful, consider the lowly mosquito. One could adjudge that it, too, "tracks" people. Leslie Vosshall has studied what makes people differently attractive to mosquitoes by bringing in hundreds of volunteers, putting them in what she calls a tropical room, and testing how many mosquitoes will fly upwind to bite an inch-wide patch of their exposed skin. (Being a volunteer at the Vosshall lab is not for the faint of heart.) Your body chemistry affects the number of bites you receive, but the number can also be reduced by simply creating a breeze. "A very helpful side effect (of ceiling fans) is that the mosquitoes get very confused," she offers: the air becomes turbulent, and though mosquitoes smell you nearby, they can't find you.*

The tracking dog cannot be so easily eluded. Unlike Paul Newman's character in *Cool Hand Luke,* who successfully confuses the bloodhounds on his scent with chili powder, pepper, and curry, prison escapees are unlikely to so easily muddle a tracker. The tracking dog's skill is not just to find the person's scent, but, perhaps more impressively, to distinguish it from the thousands of other scents gurgling and enticing his nose. A bit of pepper may cause a sneeze, but not a system collapse.

Should a track grow dim, dogs search for the "scent cone" — the invisible blooming of smelly air that spreads out from an odor source. Smell radiates from its source, getting weaker but its reach getting wider. By zigzagging, crossing perpendicular to the

* If you are, like the author, a person who not only attracts many mosquitoes, but also redirects the mosquitoes initially keen on your nearby friends and family, take these two bits of wisdom from Vosshall's work: stay downwind of the critters, and stay near fans.

odor corridor, all the while moving forward, dogs zero in on their target. Along the way they engage in a kind of on-the-fly geometry to determine when the scent has weakened sufficiently relative to their previous sniff to warrant changing direction.

Not only do we effuse a strong smell, our odor *lingers* incredibly well on things: on gauze, on paper, on plastic, on metal. We mostly touch things with our hands, and hand odor can stick around for months on some materials. Porous objects like gloves and clothing are utterly saturated with the smell of you, but even your stainless-steel watch or gold wedding ring has you on it and in its crevices. Indeed, in training and testing handlers use "scent samples" made up, simply, of a cloth that someone held in her hand for fifteen minutes — or, for the better funded handlers, by using a specialized vacuum that traps the air *above* a person and deposits it on gauze. That does it.

In researchers' parlance, our stink-wave is multilayered. We wear a primary odor all the time, part of being biological creatures. Merely sitting, doing little but turning the pages of a book, we slough off some two billion polygonal skin cells every day, along with their population of bacteria and funguses, and generate a pint of sweat (or up to five pints an *hour* should we up and exercise).* These include carbonyl com-

* One researcher fancifully quantified the smell "pollution" generated from one "standard person"

pounds like aldehydes and ketones, alkanes, and organic fatty acids — the proportion of which distinguishes us from one another. On top of that, we are fragranced with a secondary odor of where we've been and what we've eaten. Incredibly, either in an ill-conceived attempt to hide our other odor layers or in the similarly ill-conceived notion that we don't yet smell *enough,* we often add tertiary odors: perfumes, soaps, hand sanitizers, lotions, hair products, aftershaves.

The tracking dogs get all that. Lavender lotion does not put them off the particular funkfest that is your bacteria-topped, aldehydic, fatty-acid, alkane mix. And then, of course, you leave footprints. The clues in a footprint are numerous: shoe size, walker's weight (leading to a lighter or deeper track), shoe tread. The prints themselves churn up dirt and grass and leave small bits of shoe-smell in them. But also you-smell.

Don't think your smell comes through your shoes? Here's an experiment for you: Stick a pair of shoes in your gym bag, zip it up, drive home, open the bag, take your

---

(skin surface 1.8 meters; takes .7 baths a day), when sitting, to be 1 *olf.* Someone exercising may produce up to 11 olfs; a smoker, 25 olfs.

shoes out, and closet them. Return to the bag. Sniff it. Smell like "shoe"?

If you're still unconvinced, you could mimic the experiment a pair of K9 trainers did to see how much liquid — potentially carrying odor — comes through shoes. They poured water up to the shaft of a waterproofed leather hiking boot. In half a minute, water was seeping through to the outside. When walking, the boot's exterior got more sodden. A shoe emits the liquid or gas from within it, even without the wearer sweating like mad and pressing it out. And here is how the foot odor — also gaseous, even sneakier than liquid — gets out. The soles of the human foot have hundreds of eccrine glands per square centimeter — more than anywhere else on the body — which sweat. Sebaceous glands also emit odor; the fatty acids from these glands get right through shoes and socks onto the treaded surface. Your smell, via the sweat in your feet that you are constantly, normally producing, is squeezed out in every footstep.

While tracking is thus a simple matter of following smell, what is more startling is that dogs do not need to be trained to track by smell. Of course, there is plenty of smell-training to shape good tracking behavior. Early training encourages dogs to keep searching for a scent that has disappeared; handlers walk paths that point orthogonally to the wind to help catch a smell on a breeze; dogs learn to eliminate all the other cues present in the world and focus on just the one relevant right now. This training is just as much about the handler learning to help her dog do the task as it is about teaching the dog to track.

The one necessary attribute every kind of detection dog must have, by nature or training, is this: motivation. Sometimes overwhelming, house-wrecking motivation. Successful working dogs are not gung ho to track because they want to please their handlers, find the bad guy, or experience the most intense scent. No, they are gung ho to track because tracking a smell to its source gets them one thing — the thing they want more than anything: a grubby tennis ball or the business end of a shredded tug toy. Dr. Simon Gadbois, an olfactory researcher at Dalhousie University in Halifax, calls natural working dogs the "dopamine breeds" — breeds like Border collies, Jack Russell terriers, Belgian Malinois, and huskies. They are tenacious and motivated and can focus their attention on the task they need to do to get their ball.

It needn't be a purebred dog, though. A

passion for toys, a yen for fetching and tugging: these represent characteristics of a dog who can be trained to do anything in order to sate his desire to play. If a puppy is not especially interested in playing tug, a trainer may *train* him to tug — confident that if it takes, the rest of the scent-training will fall tidily into place.

To describe their work, the skilled handlers of working dogs speak almost in tautologies: "The core of searching and tracking exercises," one training handbook reads, "lies primarily in generating the desire to search and to track." Later, the authors confess, "No dog needs to learn to search." The point is this: dogs are natural searchers, hunters, trackers, smellers, pursuers. You only have to create the conditions for them to do it, and encourage it — ensure they keep liking the game. Such encouragement goes against how most of us deal with our dogs. It is almost as though most owners are assiduously teaching their dogs *not* to track. The dog who will patiently sit outside a store, waiting for you, who heels nicely and points his nose politely forward as you walk together — that dog is not destined to be a tracker. After watching tracking dogs devotedly follow their noses in search of a missing person, zigzagging in front of their handlers, seeing the familiar behavior of an owner walking her dog on leash down a city sidewalk is like watching the result of careful untracking training.

Still, training or no, watching a dog find his target through attention to invisible tracks is like gazing at a dark sky and knowing the universe sees you, but you do not see the universe. Our only access to what the dog is doing is to, well, just watch the dog. In Norway, researchers strapped a microphone on the nose of four German shepherds and set them to find the tracks of people who had secretly trodden through grass or over concrete sometime earlier that day. That the dogs could all do it (they could, in a handful of seconds) was not the issue. Instead, the researchers were investigating three distinct phases of tracking: an initial "searching" phase, ten to twenty seconds long, during which the dogs tried to find the track. Upon finding the track — about two footsteps into it — they slowed way down and entered what the researchers call the "deciding" phase. Not "deciding how I feel about this" but "deciding which way the tracks go from here." Nose in near contact with the grass or concrete, fewer than five seconds were needed for each dog to make his decision and quickly move on to the actual "tracking" phase. The head microphones captured each snort, sniffle, and sneeze, and conveyed the news that the dogs sniffed six times per second throughout. Actual breathing was only about 10 percent of what they were doing with their nose and mouth. As with other studies, the dogs were comparing the strength of the smell of each footstep — footsteps that had been laid a second apart, many minutes earlier.

MOTHER LODE

It was a garden for the blind: a constant offense to the eyes, a pleasure strong if somewhat crude to the nose. The Paul Neyron roses . . . had changed into things like flesh-colored cabbages, obscene and distilling a dense, almost indecent, scent which no French horticulturist would have dared hope for. The Prince put one under his nose and seemed to be sniffing the thigh of a dancer from the Opera. Bendicò [the dog], to whom it was also proffered, drew back in disgust and hurried off in search of healthier sensations amid dead lizards and manure.

— Giuseppe di Lampedusa, *The Leopard*

While dogs can reasonably, and with good

cheer, be trained to detect bladder cancer or diabetes, to locate bedbugs, methamphetamine, or missing persons, every dog owner knows that these quarries are not their true specialty. If there is one thing that dogs seem to have a natural propensity for, it is, instead, *finding feces.* Every dog cohabiting with a cat has deep, visceral knowledge of the placement of the litter box; outside, should a neighborhood dog have moved his bowels unattended, your dog can locate it with alacrity. An urban dog might pinpoint for you, without your even requesting that he do so, either the wandering coyote's scat or the mess left behind a tree by a local human. *What's your pleasure?*

Happily, dogs are now fulfilling their natural métier. No, sorry, Upton, not "finding feces of the homeless person behind the tree in the park." Instead, detection dogs are being put into service to find the droppings of various endangered or difficult-to-locate populations and species.

At the University of Washington, Dr. Sam Wasser is quietly heading up a dedicated facility for training and placing what he calls "professional poop chasers." Or, in the terse nomenclature of science, "scat-detection dogs." I have come to meet Wasser on an early morning at the Seattle campus, whose broad open spaces are mostly empty of students at this hour. A dense fog hides Mount Rainier, which should be proudly rising to my southeast as I head into Johnson Hall.

Though I have come to talk to Wasser about dogs, wildlife is his main concern. His detection-dog program, Conservation Canines, did not grow out of an interest in dogs. In the late '70s, Wasser worked in Mikumi National Park in Tanzania, Africa, studying yellow baboons. Like many other wildlife researchers, he was in the business of collecting information about the behavior and health of his chosen animal population — in his case, the reproductive competition between female baboons. Typically this information is gathered by stealth, trying to keep out of the animals' way and, instead, setting up camera "traps" that catch their movement, hair snags that pull hair samples (yielding DNA) as the animals wander past, or tagging or collaring animals and releasing them. Each method has shortcomings, from sampling bias (some animals are more likely to be caught than others), to the time and expense required to catch and release animals, to the stress and sometimes fatal injury from the tagging. Lures can change the target animals' behavior; trapping can injure them; even wearing a tag can fundamentally change the population dynamic.*

A better way to find out about living populations would be to treat them archaeologically: to see what they leave behind. Like bad visitors to a hotel, animals leave evidence of their activity in their wake. They graze, abandon nests, trample ground — and, most shamelessly, poop as they go. It is that poo which captured Wasser's interest.

A bit about poo. Generally, we think of excreta as worthless, without value. To call something *shit* is, pithily if not wittily, to observe that it is devoid of use or merit. Not wildlife shit. Researchers see shit and think: *gold mine.* There is information in that scat: information about health, reproductive status, what the animal is eating, and how it is feeling. The DNA allows identification of individual animals — who it is, the age and sex — and provides a way to determine relatedness among the animals. Scat samples allow researchers to reconstruct how large a species population

* This was well represented by the study of reproductive choices of zebra finches: females turned out to prefer the males not with physical extravagance or prowess, but those accidentally assigned red, not black, leg bands for tracking.

is and how widely it's ranging. Oddly, this nearly perfect collection program is such that animal researchers can gather considerable information about their subjects without ever seeing the animals.

Now Wasser's concern is the demands that human growth puts on wildlife. The scat samples allow him to suss out how the health of animal populations is changing. But the poop business began with the baboons: "I was on a mission to try to get DNA out of feces," Wasser tells me. Collecting feces could be a trial: trying to be there at the moment of production, as it were, or ferreting around in the dirt when animals had moved on. "I realized, wow, if I could figure out a better way to collect it, that would be great." At a conference on bears, where the banning of hound hunting — using hounds to find bears to shoot — was being discussed, he met a hound hunter bemoaning what he would now do with his dog. The dog, he claimed, could go forward and backward to follow bear tracks. Wasser was transfixed. "I was like, *Oh my God, that's my scat-detection dog.*"

As it turned out, he really needed an air-scenting dog, not a ground tracker like a hunting hound, and over time Wasser met and got involved with narcotics-detection-dog programs, including at McNeil Island Corrections Center in Puget Sound, to observe their dogs and training methods. "The dogs were un-be-LIEV-able," he swoons. "The prisoners would take off their clothes and go to the laundry and the dogs would just smell everyone and go *yup*" — he snaps brightly — *"that guy's got drugs"* — snap — *"that guy's got drugs"* — snap — *"and that guy's got drugs."* Wasser was convinced.

He began training dogs. His lab's first study was searching for grizzly bear in Washington State, in a place called Goat Peak. They never saw the grizzly — but the dog told them one was there. Then he performed a large study of grizzly bears in Alberta, Canada. Wasser regularly experienced the whiz-bang spectacle of his dog's nose: "I'm sitting there," he remembers, "and the dog sticks his nose in this hole — I reach down in there and here's this grizzly bear poop." Another time, they encountered a raging river, "and the dog jumps over the river and hits" — alerts on — a "tiny poop" in it.

Wasser's projects grew in size. Before long, his dogs were surveying over twenty-five hundred square kilometers in Alberta, Canada. Even in the deep snow in winter, they found thousands of scat samples from caribou, moose, and wolves, which let Wasser determine that oil-exploration activity and roads were changing the behavior of the animals. The closer the herds were to the crew activity, the higher their stress levels. He was able to tell, from scat alone, what each species was eating (wolves: mostly deer) and their seasonal variations in nutrition and stress levels. Human activity turned out to have the biggest effect on caribou populations, not the wolves that so many accused of consuming them.

Since then, Conservation Canines has trained dogs to find the scat of jaguars, tigers, and wolverines; Townsend's big-eared bat, the Sierra Nevada red fox, and the Pacific pocket mouse. Nor are the dogs mammal-centric, crossing the animal class borders to find sea turtle nests after the BP oil spill, determine the population of the northern spotted owl in northern California forest, and count the numbers of endangered Jemez Mountains salamanders, which live in dead logs and come out only one month a year during monsoons. Dogs may be trained on up to twenty species: once trained on one, adding another species comes easily. And yet they have little trouble distinguishing scat from different animals, even from closely related species, and will ignore the universe of nontarget scat around them. To alert the handler that they have found the scat, they simply sit. Not pick it up in the mouth or roll in it. Sit.

Unlike many detection programs, Wasser's dogs are mostly young mixed-breeds from shelters — because that's where dogs with excessive energy and borderline-obsessive personalities wind up. A dog with what he calls, gently, "fixation with the ball," a strong play drive, and high energy is that classically motivated dog that all programs love. "These are the kinds of dogs people think, *Oh my God, you'll never be able to control this dog,*" he says. But "they *never* run away — because you've got their ball."

Next to Wasser's computer is a calendar featuring a number of the Conservation Canines. May 2015's dog is Tucker, who has perhaps the most unlikely and spectacular job of them all. In the calendar he sits patiently on a boat with a yellow life preserver around his neck. What this sweet-faced black Labrador retriever mix does is detect the slimy scat of orcas — killer whales — who live in Puget Sound.

The resident population of orcas in the area has declined precipitously, and Wasser thought that information about their diet, hormone levels, and possible toxins could help identify why this is happening. That's in their scat — but orca scat, though smelly (think: fish) and sometimes even yellow or orange (though usually brown or green), is buoyant for only a short time and then sinks. It is not easy to find in the great wide sound. Not for Tucker. "Tucker is amazing," Wasser says. "He smells a sample over a nautical mile away and is able to track it over a fast current." Wasser is talking quickly and then stops for that fact to sink in. Though orcas are large, their scat is not highly visible. To find it before it sinks, the team has to be speedy: in choppy water, it sinks nearly at once; even in calm seas they've got only thirty minutes to find it.

The researchers take Tucker out on the boat and head in the direction that orcas have been seen, driving downwind of the animals with the wind at the boat's side. That way, they are potentially maneuvering into the scent cone (the V of the odor dispersing from the scat), perpendicular to the wind. "Tucker is asleep on the bow of the boat. But you can see his nostril going" — Wasser flares his own right nostril, which comes off like a wry grimace — "and then all of a sudden as soon as you hit it" — snap — "he is up."

"This is the most convincing of all the work that we do," he says. "There's no landmarks; the current is pushing these samples . . ." But when they enter the scent cone, Tucker stands up over the bow, his nose down, and essentially points with his nose toward the source. If the boat goes too far by it, he runs to the side of the boat. His nostrils go up and down, side to side — wet-nose rudders directing his handler how to steer the boat to reach the source.

Wasser smiles remembering how hard it was for them to see at first what Tucker was doing. In the first year of the project they never found scat. But it turned out not to be because Tucker wasn't spotting it, "but because we couldn't believe he had it that far [away]" — and so they would stop the boat and turn back. When they began baiting pie pans with samples, letting one boat follow the pan visually and the dog boat try to find it, they realized that even a mile away, Tucker was actually leading them to the pan.

At the scat, the researchers gently scoop up the sample with a little net at the end of a telescoping pole. Meanwhile, the handler gives Tucker his reward: his well-loved tennis ball. Eventually, Wasser gets his tennis ball: a well-founded result. Given the hormonal and nutritional levels that varied seasonally, he concluded that a reduction in the animals' primary prey, Chinook salmon, seemed to be driving the population's decline.

While one could, in theory, put the orca scat through the gas chromatograph to figure out what the volatile elements are that Tucker might be detecting, for Wasser, the evidence that he can find it at all is sufficient. Still, he acknowledges that there is some ambiguity about what part of the smell the dogs actually believe they should detect. For instance, he discovered that dogs trained on the sound's resident killer whale population — what Wasser called the "fish eating dogs" — were not alerting to the "transient" orcas — the ones just passing through. The transients are mammal-eaters. "We had trained him on killer whale [smell] plus fish," Wasser realized. Without the fish smell, the dog didn't interpret the smell as "orca."

But the dogs are ingenuous: they are just trying to get it right, based on what their handlers reward them for. "The handler is the hard part" of training the dogs, Wasser says. "*Way* harder than the dog." When training a dog, there comes a point when the dog-handler team moves from working with samples in a controlled setting to beginning to search in the wild. But scat in the wild will never smell just the same as the frozen, thawed, and refrozen sample from early training: it may be newer or older; the bacteria on it is different; it is from a new animal. The dog has to see what is the same between both samples, and begin alerting to that. Sometimes, though, if the team does not find a sample quickly, "You've got this highly motivated dog that wants its ball more than anything and an extremely anxious handler — because the handler loves the work and is thinking *Oh my God, I'm a failure.* And now the dog stops and checks something out and goes, *Will this work?*" Wasser looks at me from the side of his eyes, eyebrows up — the questioning, hopeful dog. "*Is this what you're after?* And then the handler goes, *Oh, well, let me see,* and the dog goes, *Oh, I guess I'm right,* and he sits down, and the handler goes, *Oh, he's right* and throws the ball. Next thing you know you've layered the wrong species onto the dog."

Even with that nose, a detection dog will look to his handler to confirm he's correct. The handler has to trust the dog enough to let him finish his job. There is not just dog anatomy involved in detection; there is human psychology.

To Wasser, there is no downside to working with dogs. Sure, it might be hard to move them around sometimes — they can't be stashed under the seat in front of you on the plane — and they are individuals, with their own temperaments and moods. But a piece of equipment — an electronic nose — would never replace the dog. "The one thing about a dog is, they improve with time," he says. And, I might add, if you're spending the day finding animal scat, there must be nothing like having an energetic, furred, devoted partner in crime.

Lyall Watson writes of a hound tracking a missing man by following a week-old trail through a bank and a grocery store, crossing traffic, and ending at the bus station. Indeed, at the very bench where the man was later determined to have briefly rested before boarding the bus. When I hear of stories like that, or of Tucker's orca-tracking, and remember, by contrast, my immediate and complete bafflement as to where my dog, Pumpernickel, had got to on that day she left the house, I think that dogs and humans have nothing in common. It seems a dog can find another animal, but it is very hard for a human animal to find a dog. But maybe I wasn't hanging out with the right human animals.

-PERSON-

Members of the Kanum-Irebe tribe in New Guinea, as described by anthropologists, have a parting ritual that may be alien to most Westerners. When two friends are saying good-bye, one reaches into the armpit of the other, "sniffs at his hand, and rubs the smell on himself," therewith eliminating the disturbing possibility that "he cannot stand his smell."

We do use each other's smell — if usually more unwittingly. At some level, we know that our own smell reflects what we eat; that we smell our age; that we smell like our smoking, drinking, and swimming habits: and so do others. Our odor reflects our mood, our health, our occupation, our medication.

Rarely do we overtly attend to or seek out these smells of others, though. But, like the Kanum-Irebe, we may actually collect others' smells — if subliminally. The now notorious research result that the menstrual cycles of women who live together spontaneously synchronize shows, in essence, a detection of others' odors, with a specific physiological result. And recently, psychologists covertly videotaped hundreds of subjects' behavior right before they took part in a (mock) experiment. The actual experiment happened in this pre-experimental time: greeted with a handshake from the researcher, people actually appeared to sniff their shaken hand shortly afterward. With same-sex researchers in particular, the subjects brought their hands near or to their noses after shaking hands, and sniffed. It was as though they were sampling the smell: what the researchers called "chemo-investigation of conspecifics." Smelling you. Certainly most of this chemo-investigation is unconscious.

Can people intentionally track other people or other animals by smell? I resolved to find out.

The porcupine is a messy animal. Compact, his body covered with tens of thousands of hairs modified into barbed quills — which, famously, a feisty porcupine can release and drive into the flesh of anyone foolish enough to come near — he is well armored. Before he does that, though, the porcupine will clack his teeth at a potential predator, and release a penetrating odor from glands on his back. He would rather *not* get into a scuffle.

It is, perhaps, the many layers of defense that the porcupine, *Erethizon dorsatum,** has, that allow him to be so, so slovenly. The porcupine pees where he pleases, often while walking along, not bothering to slow or stop. The footprints of a porcupine may be peppered with bits of dirt or defecation, as he sleeps in a den piled with his own excreta. Unsurprisingly, he does not groom himself. When animals groom, they are doing so less out of a sense of the importance of cleanliness than out of a fear of predators scenting them or pests injuring them, or out of an interest in mating with other like-minded animals. The porcupine handles predators with aplomb, seems to have a hearty immune system, and has figured out the mating game even with all those quills. So he gets to be unkempt with impunity.

* Which translates roughly as "animal with the vexing back."

His grubby manner is how it came to pass that I could tell a porcupine was near us in the forest, one cold January day. Reader, I caught a waft of porcupine urine.

The day had begun eight hours before, under three layers of night that gradually yielded to day along my drive from New York City to western Massachusetts. I had risen early to go animal tracking. If there was anyone who could rival the scat-detection dogs, it was scat-detecting animal trackers. I was headed two highways and a string of desultory industrial towns away from where I awoke, where I planned to meet Charley Eiseman and Noah Charney, naturalists with an ecological, philosophical, and purely hedonic interest in finding the tracks and sign (read: often scat) of animals.

In the dark of the city, I kissed the top of my son's head* and headed outside to my car. I thought to sniff the city's nighttime air, knowing that I would soon have a contrast in the bright, cold western Massachusetts climate. I had to work my nostrils a dozen times to notice anything at all. The city smelled ashy — dusty, maybe. It smelled like . . . absence.

* smell: sweet, sweaty hay.

■ ■ ■ ■

In the excitement over dog-tracking ability, one might forget that humans have a long and storied history as trackers themselves. Certainly, as an omnivorous species, and before being an armed species, we humans found and trapped animals for food simply by knowing their habits: where they lived, what they ate, when to find them out and about and vulnerable. These days, apart from sport hunters, people rarely track animals to find dinner; instead, the extant "animal trackers" are hunting animals largely to photograph them, to survey the population, or to feed their own inquisitive spirit. In this way, contemporary animal tracking is a curiosity rather than a necessity. So, too, for the kind of "tracking" a contemporary American pet dog might do: for the great majority of twenty-first-century dogs, following their nose is less about finding a mate or dinner or minding territories than it is simply noticing what's around them.

The phrase *animal tracking* gives the impression that the end result is finding an animal, but this is not so. The most likely outcome is that one will not see the animal at all. Nearly every wild animal will notice a human wandering in its midst well before the human notices the animal in his. Instead, tracking is about discovering the various indications that the animal has been by. The classic sign is an animal's tracks, or footprints. Since we were headed out into a mostly untrammeled forest, where snow had fallen a few days before, there would be clear, discernible animal footprinting for us on this day. Mud is also a fine substance for catching the footfall of the passing coyote, turkey, or moose. Most terrain, though, does not capture the forest's populations' light footsteps (such that humans can perceive it), and most weather — rain, continuous snow, wind — is well designed to wash any traces away. A tracker often doesn't have, and doesn't need, actual *tracks* to track an animal. Any forest is, down to each hillock, to every tree and shrub, wearing evidence of the animals that live there. All of our senses, and all sorts of evidence, are used to trace the passage of an animal — including their smell. Deer smell. Squirrels smell. Bears, bobcats, foxes, and moose smell.*

* In her wonderful volume *A Nosegay,* Lara Feigel quotes the story told by Edmund Snow Carpenter in *Eskimo Realities* (1973). He recorded this

What might a moose smell like, you ask?

"Oak-y," says Noah Charney, as, nose in the air, he cuts a corner on our path and wanders among a grove of trees. "I thought I smelled a moose." He pauses. "But it [both the moose and its smell] is gone now." Watching Charney nose-hunting for moose is like watching an art appraiser eyeballing the ostensive masterpiece placed before him: while one can see him *looking,* one has no idea what he is *looking at.*

What Charney, and every animal tracker, does is learn to embody the animal he is tracking. To understand where the animal might be, one has to *think* like that animal — to imagine its life. The skill of tracking has been compared to learning to read: only one reads not the grade school primer, but the forest. As a book's hieroglyphics morph into legible prose to the young reader, a wilderness begins to look the way it might to the animals who live in it. What bush would thoroughly conceal you, if you were a rabbit? What manner of tree presents the

---

exchange between an Inuit woman (W) and an anthropologist (A): W: "Do we smell?" / A: "Yes." / W: "Does the odour offend you?" / A: "Yes." / W: "You smell and it's offensive to us. We wondered if we smelled and if it offended you."

perfect height and visibility for a bear to announce her presence by rubbing and clawing along it? Which secret hollow appeals to a flying squirrel? What time of day might the nearsighted possum feel safe foraging with her pink grubs of pups? Once one sees the world like the animal, becoming aware of oneself *in that space,* just as the animal is in that space, one finds traces of the animal.

Of the animal tracker's bag of tricks, it is the tracking of smells that I am after, of course: the odorous sign of an animal's passage. Heading up north in my sealed-off car, I have only a vague idea exactly what those signs might be. When I nuzzle my nose in the scruff of Finnegan's neck, I surely recognize his smell. I have learned that when guests come to our door, they note the distinct odor of "dog"; we, who inhabit the house, have become so accustomed to the fog of dog that we cannot smell what they smell.* Surely Finn wears, or emits, other odors, but even in our close compan-

* This phenomenon is called habituation: no longer noticing an odor after repeated exposure. While it is similar to adaptation, the latter operates at the level of the receptor cells; habituation is a result of the *brain* no longer bothering to notice a smell.

ionship I have never thought to loiter on them. This is about to change.

It is still the early hours of dawn when I pull up to the tracking-school classroom building. It is deeply cold: eight degrees Fahrenheit. The deepness of the chill is heightened by the stillness of the morning. Cold, calm winter days are good for ground tracking, the animal-tracking manuals say, presumably because, apart from the recent animal smells, all the other volatile odors of the world — from tree, plant, and earth — are quietly asleep beneath the cover of snow. The warm touch of an animal is a beacon of odor in a cold scene; this warmth even volatilizes whatever it touches, creating bubbles of odor in a barren landscape.

I want to properly prepare my nose. Admittedly, this preparation is pretty straightforward: I blow. Make room, remnant city and car smells: I need to smell coyote! For good measure, I've brought along a nasal steroid recently prescribed to clear some congestion plugging my ears. It can't hurt, I figure, to be a nose on steroids. "If I wanted to improve my [scent] detection," Stuart Firestein had admitted to me, "I would use steroids." He was not making a doctor's recommendation, just an observation based on his own past experience. Due to an oral surgery, he needed a nasal steroid temporarily. "I smelled things I had never smelled before," he waxed rhapsodically.

The spray is like a sudden injection of a lily into my nose, petals, pistils, pollen, and all. As I inhale, I can feel my nasal passages opening, stretching their arms and widening their eyes to the daylight. If there's any time that my nose might be a super-nose, it is now.

Inside, I find Charney and Eiseman in a small, busy classroom; Charney lifts his eyes in the briefest of greetings. They both share the same unruffled manner. Both are dressed comfortably in muted layers. Two packs are laid out across the chairs, spilling supplies. Their equipment has the specificness of those who are prepared to be outside for long periods, but who do not like to be over-prepared.

Shortly, a half dozen tracking students stumble in, all heavily dressed for the day's expedition. "The bulk of our time will be spent in the field visiting local habitats," the course description had warned, telling students to "be prepared for long days in cold, wet, and strenuous conditions." Hand warmers are in evidence.

Charney approaches and wordlessly hands

me something. It is an eighty-page tan Steno Notes notebook, its pages plumped with wear and water. His tracking notebook: the repository of notes and ephemera from past expeditions. "Noah's scratch-and-sniff book," Eiseman calls it, because it contains physical leavings and evidence in the form of fur, leaves, sticks — and in the form of tissue paper sodden with urine.

I handle it gently, as though it were an ancient scroll. On one page, a long, tassel-like hair is taped on the page. "Moose" is scribbled by it. On another, "Fox, 12/27/01." Some pages hold a series of twigs with incisor marks, or clumps of fur that seem not to have fallen from their bearers naturally. There is sign of rabbit, caribou, bobcat, possum, bison. "3/03 Beaver" holds a patch of beaver hair; under it, "Coyote," and a tissue once moistened with, presumably, coyote urine. I wonder whether that coyote and that beaver met on that day; I sniff the page. Riffling the pages through to another tissue, I sniff that one, too. Apart from the surprisingly clear "Steno notebook" smell, I can detect that there are *some kinds of smells,* but they betoken nothing.

Jammed between two of the final pages is a sample with High Security, for the field: a crumpled tear of aluminum foil is wrapped around a ziplock bag — itself securing a folded tissue. A note on the bag reads, "Red fox, 1/9/03." Charney carefully extracts the sample, which has not seen fresh air in twelve years and four days. He barely touches it to his nose, with the gentleness of a man sniffing a woman's handkerchief. *"Oh!"* He has a physical response to the odor, pulling his head back and striding across the room away from it. I take a turn with the tissue. My nose catches something sweet, *animaly.* Eiseman leans toward it. "Skunky!" he says at once, smiling.

Charney nods. The simple syntax of animal-urine odors seems to be: name of animal evoked (skunk) or food source (oak), plus punctuation to reflect the intensity of the odor. *Skunky!* works for both skunk and red fox. *Musty,* for a bear. To both Charney and Eiseman, this odor is exclamation-point loud — and patently from the elderly bladder of a red fox.

Eyeing the tissue, I feel a bit like a color-blind person in a room of rainbows: I sense there is something to be seen, but my searching eyes can't find it. Charney hands the tissue to the students, hesitantly poised at the table in full winter garb. Each willingly brings it to her nose and sniffs it. Last week, before the class, such an offer would certainly not have been accepted by any of them — or, I daresay, by many outside of this class.

Charney is the exception. "I've always smelled everything," he tells me, "for as long as I remember." He tells a story of living temporarily in a wigwam in the woods while he took classes by day. Returning to the forest in the depths of night, he had to navigate by smell. "One time I did wake up and I didn't know where I was," he admits. Usually, though, he found his way by following his nose. Remembering my lack of familiarity with the smells of my own block after smell-walking through Brooklyn, I comment that this may be atypical. He shrugs it off. "I'm always surprised [at the things] that people can't smell." He laughs, a bit vexed. "I'm like, *Why not?*"

That Charney does not see his exceptionalism is obvious in his description of meeting his wife for the first time. What he remembers, he says, is that "when we first started dating, she smelled like Coca-Cola." He recalls his childhood as redolent of his toys: "The plastic tires on those little plastic toys? They smell this really strong vinyl-y kind of smell . . . And the *whole reason* to play with Play-Doh is the *smell* of Play-Doh." As with perfumers and wine experts, when he describes the vinyl smell of toy car tires or his wife's cola ambience, he is not just remembering the smell abstractly: he is smelling it in his head.

This ability to invoke odors, to smell-daydream, as it were, is a clear marker of those who know how to use their noses, and those who do not. It marks attention, the brain's willingness to loiter on an experience, which is later conjured up. Some people experience smells in dreams — though even normal smellers seem to have their dreams affected emotionally by odors in their environment.

Though odors have emotional content, for Charney, they are simply a part of the scene, and a customary note of his daily experience. He has odor preferences, as we all do; where he differs is in his approach to the vast expanse of odors that are not naturally pleasing. "There are smells I hate. But I still like to smell them — because it's like, *Ooh, that's horrible.*" He grins. What kind of things, I wonder. "You know, vomity kinds of things," he replies.

Oh.

Everyone piles into a van and we head farther from highway and habitation. Our destination is the northern end of the Quabbin Reservoir, a water source for Boston surrounded by state forest. Charney drives and Eiseman plays bird songs on the radio. Apart from being in the outdoors, this is clearly their element, listening and smiling and nodding in recognition of whoever sings *tuh-wheee!*

After an hour we park and tumble out of the van. The forest at the road's edge is a wall of undergrowth and fallen trees and branches; there is no obvious path. Charney simply pushes a branch aside and walks in. After a beat, we all follow him. The woods immediately swallow us. There are occasional animal "runs" that serve as makeshift paths, but we mostly bushwhack through tall, tight reeds and bushes, forging serpentine routes between low branches of close-growing hemlock and white pine.

A dozen minutes in, the only signs I can detect of where we might be are our own footprints in the snow and the ascending late-morning sun from the southeast. I am engrossed in the activity of minding where my foot is going to step, and notice little else about our surroundings. But the trackers have their heads up: while I walk, they investigate.

Charney stops. "Do you see this? Have a look." And he walks off, leaving us to examine the tree he's gestured toward. On first examination, it is, unquestionably, a tree. Nothing remarkable about that. Then our eyes shift. It is damaged — there is a tear, or some holes. Peering closely there is a single hair — a hair! — sticking horizontally out of the bark. It feels like a miracle to see it. Charney returns. "Did you find the hair?" It had popped out at him, bogglingly, while he was striding by. He then diagnoses the scene: this is a tupelo tree, in a bit of clearing, on a small hillock over what are now-frozen wetlands. Nearby are a handful of larger tupelos.

That is what the tracker sees first. They see that setting as a great fire hydrant on which a female bear might leave her mark. The trunk damage: bite marks and scratches from her broad claws — both ways to leave her scent behind as a warning to competitors. The hair? From rubbing her back against the tree's trunk. Those larger tupelos could be "nursing trees" — up which a mother bear can safely stash her cubs while she forages for the tree's coveted berries.

I peek around me carefully. No bears in sight. Still, the air is charged with the scene evoked by the traces left behind.

As we carry on, we slowly but surely collect animal sign: the track of a flying squirrel, which begins suddenly (its landing site) and quickens to the safety of a tunnel at the base of a nearby tree; cottontail rabbit; deer so frequent they are unremarked upon. Sign can be claw marks or shed hairs; traveling paths forged through close undergrowth; depressions in grass or fur where an animal lay and nested. And excretions: droppings, secretions, other often mephitic exudations. "There is a whole novel in an owl's pellet," a tracking guide book rhapsodizes. The book is chock-filled with full-color photographs of scat and mucuslike blobs, centered respectfully in their frames. The novels written by excreta are portraits of the excreter: his species and sex, health and diet, preoccupations and companions, and daily routine. A coyote's scat is filled with fur or hair, contrasting it with dogs', rife with the grains that make up packaged dry dog food. A pile of lake muck, topped with a yellow-orange smudge, is sign of a beaver feeling territorial.

The human tracker may constitutionally be directed by his eyes, but his nose brings more information still. An experienced tracker like Charney can air-scent, if the air is sufficiently humid and the wind is at his face. But to do as the animals do, it is also useful to get at animal-height: the "stick your nose right in it" method. The general instruction to "get down on your knees, put your nose as close as possible . . ." (as one tracking bible says) is anathema to most people. But Charney regularly drops on all fours, his face a half inch from a moss-covered stump or tree trunk. He races ahead into a clearing; when we catch up with him, he is getting up from sniffing a wide stump. Everyone takes a turn on their hands and knees, nose up to the broad vertical shelf of the trunk. "Remember to blow some warm air on it so the smells come up," he advises. Whatever odor is on that trunk, it needs the warmth from our lungs to make it volatile, to evaporate into the air and be captured by a sniff. Similarly, the lightest of rains can "breathe new life" into otherwise quiescent odorants on the ground. I warm the trunk with a breath, close my eyes, and sniff. I smell a mustiness, reminiscent of basement. Eiseman is smiling: he already knows what animal has been here, just from the context. He confirms it with his nose. "*Cats* in a basement," he corrects me. Here, *bobcat.* What I recognize as *basement* is actually the smell that my past cats have left there.

There are no footsteps, no other obvious sign of an animal being by. So how did Charney know to sniff *just here,* then? "You have to think about whatever is prominent for the animal. I just look around for all the places that might be interesting to mark," if you're an animal that urine-marks — a single item sticking out in an otherwise undifferentiated scene, say. And "context-wise, this area" — with a stump alone in a small clearing — "looks good" to attract a marking feline or canid. And then you smell.

Bobcats are so specific in their marking habits that animal trackers can be likewise specific in where to look for their marks. "Urine deposits are usually eight to nineteen inches above the ground," one book instructs. "The most common scent post is a short, decaying stump, usually not more than six inches in diameter and four and a half inches high. If the stump is leaning, [bobcats] usually deposit urine on the underside where it is better shielded from the elements." We had aimed about fourteen inches above the ground, on a decaying stump with a lean.

Indeed, a good amount of smell-tracking lies in using vision, too. That's fair. Dogs, certainly, use their vision. Their being olfactory creatures does not preclude their *looking:* they don't find another dog's rump only by air-scenting. They may see the body first and follow it to its derriere: *then* they sniff.

The whole point of scent-marking on a fireplug is that it is easy for the next sniffer to locate. The sight of a stump in a forest clearing presumably does for the forest-dwelling animal what the sight of a fireplug does to an urban dog. And I, at the other end of that dog's leash, see it, too, and know where he will lead me. That behatted, upright fireplug planted right along a boring flat stretch of sidewalk: that's a fine place to put a mark. Footprints — another visual clue — might lead one to a likely smell site. And so it is with animal tracking.

On top of the contextual clues, we could eliminate the possibility of the marker being a canid even before putting nose to stump. If it had been one of the local wild canids, the red fox, "we'd be able to smell it from here," Charney says, stepping back eight paces. "A red fox really assaults your nose." Trackers tend to call red fox urine "skunk-like," much stronger than domestic dog, coyote, or gray fox urine. Unlike the relatively modest — if not exactly pleasant — ammonia of the bobcat urine.

We eat lunch on the frozen wetlands, glad for the cloudless sky. I turn my face to the sun and curl my fingers into fists after quickly downing my thermos of tea and sandwiches. For a while, everything will smell to me like peanut butter.

When we continue on, conversation has dimmed, but the sound of our footsteps obliterates any other sound. We are not quiet animal trackers. At a broad trail I hang back from the group to listen to the forest. The day has warmed under the steady sun, prompting the crystalline sound of tree ice showering onto lower branches. I take sniff-samples of the air; I sidle up to trees and surreptitiously give a sniff. I feel the cold in my nose and can almost taste the clarity of the country air.

Ahead of me, footprints form a track. To my naive eyes, it appears to be bisected by one or more indistinct sets of tracks running across it. I follow the track down off the path and into an area of low brush. There, near a little spring of a hemlock, is a dribbling trail of light yellow. I stop.

Now, I've seen a lot of pee. I live with dogs, after all: which is to say, I spend some time every day overseeing my dogs' urination. Indeed, one of the strangest things about being an urban dog owner is that we owners have great insight and attention to the evacuations of our charges. We know their methods (leg-lift, squat, squat-and-scoot, sniff-and-circle), their preferred locations (in the curb, on the hydrant, with grass or leaf litter underfoot), and even the expected quantity of output. I'd hazard that, accidentally, we are familiar with the smells of our dogs' urine, just as parents become intimately acquainted with the sweet odor of their own newborns' blissful defecations and micturitions, simply by constant exposure.

So I barely hesitate before kneeling in the snow, crouching low, and sticking my nose ever closer to the yellowed snow. And I smell . . . porcupine.

Porcupine pee smells — there can be no question — *piney.* In the winter, the animals subsist mostly on the bark, leaves, and buds of pine trees. Scattered hemlock branches on the ground, incised at 45 degrees to vertical, are evidence of the foraging of this particular porcupine. They climb trees, gripping with their impressive claws, and make their way out to the ends of branches, holding the newest of the new growth. This arboreal habit, while providing them with ample calories, is also sometimes responsible for their ends: a porcupine who falls off a tree lands on many sharp, impaling quills — his own.

I stand up, beaming, and glance at Eiseman. "That might be my favorite animal-pee smell." He smiles. Well, if you're starting with the "smell of urine," I have to agree: it isn't too bad. Woodsy and strong, but not overpowering; I've smelled much worse in NYC taxi air fresheners. We follow the footprints leading away from my find: the back feet come down slightly ahead of where the front feet lay — an "indirect register." The toes point out, befitting a waddling, short-legged, chubby-bodied animal. Aside the toe holes are lightly rendered lines drawn by the quills grazing the snow; within them are small chunks of dirt and forest debris. The profligate peeing, the messy trail, the distinct shuffle: porcupine. "Porcupines and red foxes are the animals that I will often smell a hundred feet away," Eiseman adds. "I only sometimes *see* the animal."

Do their tracks themselves smell? "I'm sure that to a fisher" — their most common and successful predator — "they smell." To a porcupine, too, perhaps. Nosing a four-toed footprint,* I confirm that I have neither porcupine nor fisher genes.

We follow the trail to its end — a den with remnants of a stone wall on three sides — and peer in hopefully. This might have been

* Its front foot; a porcupine's back feet have five toes.

an icehouse a century ago, before the state flooded the four towns in the area and created the nearby reservoir. Instead of porcupine or porcupette, there is a towering pile of pellets, many feet high, in either corner: insulation for a cold winter. As the wind turns, the fecal smell floods the air. If the quills aren't deterrent enough, this could prevent other animals from trying to steal the den.

Midafternoon. We stop at a frozen beach, taking in some more of the sun's warmth and gazing out at the long expanse of iced lake at our toes. Twenty feet out on the ice, the trackers are looking down.

"Hey, Alexandra, there's coyote pee!"

Rarely has this utterance inspired excitement, anticipation, or thrill. Or, I'd guess, a quick jog toward the informant — as I set to do. As I approach, I can see that Charney's nose is inclined toward a small twig frozen into the ground, protruding only a few inches. A single dried leaf shivers from its top, and an expanse of moss is wrapped around it. At a few inches tall, it isn't much to look at — but it is the tallest thing around. Looking out to the ice and around my feet on the beach, numerous trails clearly converge on this local landmark. It's perfect for a coyote, who prefers to mark highly undistinguished items. The coyote passes by the towering oak without breaking his stride; he snubs the ancient hemlock, the prehistoric rock outcropping. No, it's "small bushes, tree trunks, rocks, blocks of ice, mounds of snow" that shine brightly in the coyote's eyes. This twig is just right: though only noticeable to the human when she trips over it, it is nonetheless a single bump in a sea of flatness. The tracks are about a foot from the twig, indicating that this was probably a male, who could reach the twig with a spray under a lifted leg, rather than a squatting female.

I crouch and hover over the stick. There is a splash of yellow on the snow, a more profligate leaving than the single droplets doled out in the forest. I sniff. Smoky, earthy. Incredibly strong.

"Doesn't have the skunkiness of a fox," Charney offers. "Pure ammonia," Eiseman says. "A lot milder than fox." Charney leans in and takes another whiff — "Whoa!" — and he keels over, flat on his back, in mock (partially real) horror. "There might be a domestic dog involved here." Meaning, this smells like an animal that is eating a different sort of diet (one coming out of bags and off of kitchen counters).

Indeed, though the large, loping footprints converging on the stick look like coyote, some tracks are much more disorderly than the others. The dog's gait is, in fact, considered "sloppy." The dog indirect registers — leaving twice the footprints. His nails make distinct nail-marks, spreading out from the pad, unlike the tucked-in pithiness of the coyote print. The dog's course is meandering. "You just don't see wild animals making some of these tracks," Eiseman explains, "because they're trying to conserve energy." Coyotes' paths are clean, straight vectors from A to B; they "direct register," back feet stepping in the front feet's track. They cannot afford to explore every new thing, every which way. This wandering track is from someone that's getting fed.

We linger a bit at this perfect scene. Not an animal in sight, but sign of animal all around.

The sun's quickening descent tells us that it's time to find our way out of the forest. Though we have wandered for hours, over many miles, my guides seem to know just where we are and beeline for the van. "Every forest has a smell," Charney suggests as we walk. "This one smells like the leaf litter when there is no snow," says Eiseman, "of the blueberry and gorseberry undergrowth." If they were dropped into this scene, blindfolded, with no wind on their faces or sound in their ears to go on, they could still identify where they were.

Eyes open, ears on alert, nose exercised, I try to take it in. Back in the city, hours later, I can distinctly smell my urban forest. In the first half of the twentieth century, the MTA — the Metropolitan Transportation Authority, responsible for running the city's subway system — employed James Patrick "Smelly" Kelly to walk the tracks and sniff out any gas or water leaks. His skill was legion. Once, presented with the complaint of a bad odor beneath the subway a block off of Times Square, he arrived at the site to find distinct odor of elephant. Indeed, it turned out that the Hippodrome, which housed many a traveling circus, had been at Sixth Avenue and Forty-third Street, right above the station. Apparently the circus elephant dung had been left behind, and its smell was reanimated by a nearby water leak.

I sniff for, and do not find, elephant in the city air. Often the city's smells are noticed only when foul, but now they feel neutral, informative. I breathe them in with satisfaction. But I do mourn the lack of a certain pure, icy northern snowflake in my nose.

10:
Civet Cats and Wet Dogs

> And every day, as he came near to it, he would lift his small pointed nose high in the air and sniff the wonderful sweet smell of melting chocolate. Sometimes, he would stand motionless outside the gates for several minutes on end, taking deep swallowing breaths as though he were trying to *eat* the smell itself.
>
> — Roald Dahl,
> *Charlie and the Chocolate Factory*

It is May 19, and the state of Washington is gray and evergreen. The highway out of Seattle is a course of improbably-tall and improbably-taller trees, until finally, far enough north, they tickle the undersides of the clouds.

I'm traveling, and missing my own dogs. Were I home now, I'd be snapping on leashes and heading for the park, the dogs jostling each other to be the first out the door, out the elevator door, out the building door. As I drive I chuckle remembering a typical Monday morning tour through our local park, during which Finnegan locates the weekend's best picnicking spots and scours the grass for any edible contributions. Today's outing would certainly be of interest to him — and to any dog who tends to arrive unbidden from across the room to snarf the tiniest morsel of cheese left after sandwich-eating.

I'm going hunting for truffles.

-Dog-

It seems a shame that dogs aren't employed to work at finding smidgens of cheddar. One might charitably call the yellow Lab who steals bagfuls of bagels from unsuspecting picnickers a "bagel-detection dog," but rarely is one in need of this particular kind of breaded acuity. For a short time when our new cat was still introducing herself to our home hiding-places first, we employed Finnegan as a kitten-detection dog. Now that the kitten sleeps on my face, the dog's services are no longer needed. Alas, there is no employment for hot dog–detection dogs, awesome though most untrained dogs would be at this work.

Enter truffles. Truffle-detection may be

the perfect convergence of dogs' desires and our own. For to dogs, truffles smell fantastic; and to humans, whether or not you find truffles fantastic-smelling, they smell like gold. A golf ball–sized Périgord black truffle from Italy or France retails, at last check, for over one hundred dollars; enormous 2.5-pound truffles have sold for hundreds of thousands of dollars. Truffles are "hypogeous fruit bodies of the ascomycetes Tuber spp." — that is, an underground mushroom, one that has its spores in its belly. It's unusual among mushrooms in that it forms a symbiotic relationship with the trees to which it attaches itself: the tree gives the truffle a landing spot, around their tendrilly roots, and carbohydrates for growth. But the truffle, extending its own reach, seems to gather nutrients for the tree. Truffles don't grow without trees, and trees grow better when entruffled.

An industry has formed around unearthing these delicacies. They are often only inches underground — even a centimeter in some cases — but are not easily spotted. So how to find them? You could dig willy-nilly, but given the rarity of truffles, therein lies madness — hours of digging and little to no yield. Some rake the ground, hoping a truffle will be pulled out on a prong, but this reveals more immature, untasty specimens than not, and damages the forest floor to boot.

Truffles have a related problem. As a spore-driven reproductive system, they need to find their way to the light to send their spores far and wide. So the truffle has figured it out. When it matures, it lets out a hair-raising odor. The odor draws animals to pursue and dig up the truffle, consume it, and thereby spread its spores wherever the animal travels and defecates. Scientists have identified some likely culprits for the particular *eau de truffle:* it includes dimethyl sulfide (think boiled cabbage) and androstenone (a hormone also in our sweat).

So-called truffle flies love the smell — and you could try to track a scrum of truffle flies as they tumble over each other and dive toward it — but nicer might be a quadrupedal vertebrate. One who likes smelly stuff.

Ah, there's a lead: androstenone, which we met before, is a pig and boar pheromone, and whether or not pigs find truffles erotic-smelling, they have no trouble locating the source of the odor, even underground. As a result, pigs have been used to hunt truffles in Europe for centuries — their snouts turned down, already half in the mud and dirt, unwavering in their course. This same smell has an unusual human history: it is one of the rare cases in which an olfactory receptor has been identified (named, unpoetically, OR7D4) for which variations in the gene responsible for the receptor lead to different reported experiences of the smell. Some people find the smell to be like urine or a sweaty sock; others find it pleasant, even flowery. Other people (including yours truly) appear to have a variation in the gene such that they cannot detect it at all.

Those people would *not* be good truffle pigs. But sometimes truffle pigs are not good truffle pigs either: they find the truffles naturally, but also love to eat them. Points for efficiency: they are discoverer and consumer at once. Enter the truffle dogs. While dogs need to be trained to care about the particular smell of truffles, they can easily find them for their owners or handlers once they do. And they won't eat their find. Not all of the time, anyway.

I pull my rental car off the highway into a nondescript rest stop where we have agreed to meet. I see the trainers first. Their dogs — and they have brought four of them — lie quietly in separate crates in the backs of their cars. Alana McGee hops out of her car and presents me with a nondisclosure form I must sign. She is dressed in a Truffle Dog

Company jacket — the company she co-founded — with a logo of a dog digging himself into a hole; her hair is pulled back casually into a single ponytail. From a second car, cofounder Kristin Rosenbach greets me guardedly, an occasional smile transforming her face.

The nondisclosure form is typical of the mystique that surrounds truffle hunting: see the possible prices for the fungus. In this case, McGee is hunting on private land, where she has been granted special permission to sniff and dig. While the presence of truffles in North America is not widely known, in the areas such as Oregon where truffles have been found, an identified hunting site will attract opportunistic hunters, who ransack the forest using destructive methods.

After signing and peeking in at the dogs, we all drive separately to their first location. The roads get narrower until we turn at a former game trail with no identifying signs and park. As she gets her gear together, McGee describes how she got interested in truffles. She is a certified dog trainer, but only began training truffle dogs in earnest after visiting Italy and watching the *tartufaro* work with *i cani da tartufo.* Returning to Washington, she realized that she was living in a potentially rich environment for truffles; indeed, Oregon has truffles, and the habitat is very similar one state north. With the help of another truffle dog trainer in Vancouver, Kelly Slocum, she trained her black Lab mix Duff on the scent. Now all she needed was the land to search on.

"I just started approaching timber companies," she remembers. "Wow — they are not used to getting approached like that" — by a young woman who wants to nose around the land with a truffle dog. With each, she simply suggested that she and Duff explore some of their land, with the notion that there may be mutual economic benefit should she find a rich vein of truffles. A number of companies were receptive. She explored, following foresters' leads as to where she might find success — but also looked for characteristic features of a forest that make it a likely truffle forest. This includes a wide range of ingredients: the age of the forest, types of trees, the level of disturbance or clearing, the amount of sunlight, the soil consistency. Possible trees include pines, oaks, birches, willows: all can host the fungi that are truffles. In the Northwest, Douglas fir, Scotch pine, and noble fir are likely — and common — hosts. Black truffles "really like" ferns, as they keep the ground moist: so she would look for fern undergrowth. Given a plausible forest, one then has to get dirty. "First look around on the forest floor," one truffling field guide recommends. "Have the local experts — the small animals" — like squirrels, mice, even coyote or raccoon — "been creating little digs in the earth?" All these animals look for truffles as part of their diet. One simply needs to dig gently for a few inches; truffles are often just under the top layer of the forest floor. After some searching, and despite some harvest-killing cold snaps for a few years, McGee pinpointed a handful of good sites. "I still have keys to several of the blocks. They would pretty much give me free rein as long as I reported back on what we were finding."

So grew her business. Rosenbach, who has a background in human education, found her by googling *Washington state truffles* after her dog Callie dug up a truffle unbidden while they were walking together in the forest. They soon joined forces.

At her name, Callie, a Border collie with a becoming half-white-half-black face, perks up. Rosenbach has opened the back of her vehicle, but it is Cash, a shelty, who is going to hunt first. He barks — *arr-rup!* — lifting himself with each exclamation. He has runway-worthy long hair and broad furry feet.

"We might find a truffle on the side of road," McGee says. "We work here." Since they both hunt truffles and train students at this site, there is always the possibility that a truffle falls out of someone's pocket. If so, her dog will find it.

Lolo jumps out of McGee's car. She wears a brown-and-white fur coat, her hair in tight curls, and a fitted jacket to protect her from errant twigs, barbed wire, and tossed bottles one might encounter in the forest. Lolo is a Lagotto Romagnolo, a breed you have likely never seen around the neighborhood. Bred in Italy specifically for truffle hunting, the Lagotto has been slow to catch on in the United States and remains a distinct minority. It's hard to see why: the dog is personable, engaged, *sportif.* Her fur is captivatingly soft and dense, the teddy bear of imagination, good for childhood bedtime snuggling.

There is a certain resemblance between dog and owner, as so often happens. While the dog alone has the curly coat and the lolling tongue, Lolo is, like her handler McGee, compact and athletic. Lolo's head is a darker tone than her body hair; it's hard to tell if this is from the holes she winds up plunging her head in, or if it's her natural coloration. Ringlets of fur hang down over her eyes — but she doesn't much need to see. She is on the scent.

Training a dog is what McGee describes as "reward-based detective work," an exercise in expanding and contracting the challenge to dogs. Once dogs learn the game of searching and rewarding and are imprinted on a scent, the process can begin in earnest. An early training situation might involve setting up a simple hunt inside: a truffle oil–infused cotton ball in a box. Once a dog reaches a level of success or comfort there, the scenario is complicated — the target is hidden up high, or underneath other objects — or the criteria for success are increased — the target odor is weaker; a subtler alert is accepted. From there, one might move outside — expanding the challenge, but reducing the criteria for success — contracting it. All along, the challenge is to teach the dogs what the game is — finding the scent, not "finding boxes" or "barking" — and teach the handlers to be sensitive to their dogs.

The forest of towering fir trees smells like growth, fertility, promise. Moss grows everywhere. Pine needles cover the floor; ferns, some still in their fiddleheads, are profligately scattered. The dogs dive in, trailing long, brightly colored nylon leashes. "Yeah," McGee says, speaking to the unasked question about their motivation. "We have to use games to get them *out* of a truffle forest." She walks quickly after them. Once we have stepped inside the forest, the road is invisible behind us. It is mesmerizingly silent but for our own sounds: our steps crunch small twigs underfoot. A pouch holding a shovel and a spoon from McGee's belt clangs against her leg.

McGee starts the search at the bottom of an incline, heading generally upward, because odors tend to cascade downhill. To check airflow McGee may bring matches, watching the smoke as they are struck, or a device called a wizard stick, which gives off a visible vapor. Just as indoors at the Working Dog Center, making air visible helps us nose-poor creatures help the nose-rich.

Looking up at the hill ahead and catching no whiff of truffle, I imagine the forest's dark depths as the arboreal haystack into which we will mine for a singularly elusive needle. The search seems impossible: the forest surrounds us, uniform in its ambivalence to our presence. But almost at once my analogy is proven utterly inapt. In the dog's hands, as it were, the needles are rife:

we can't but turn without being pricked by one.

"Stay with me!" says McGee to Lolo, hurrying a bit. "Lolo, good girl. You can do it. Where is it? Here? Show me?" She takes out her red-handled spade and jogs over to Lolo, nosing deep into the fern. Twenty yards away, Cash gives a little sneeze. He is nose-bent to the forest floor — what they call *duff* — and barks.

Within minutes, both dogs have found truffles. McGee and Rosenbach quickly dig them out, cache them, and pat the duff back into place. Lolo is already out of sight. McGee keeps up a constant one-sided conversation with her. "This way, please! Let's go! Thank you . . . come here, love. Can you show me? Good girl."

What was her alert? I wonder, having seen nothing more than a dog pausing and then taking off into a run again. "She digs," McGee admits. "I'm trying to work on nose touches" — pointing the truffle out by touching it with her nose. "Dogs tend to offer alert behaviors organically" — behaviors like sitting, lying, even a pause — "and we tend to go with those. And there are those we teach, like nose touches." Rosenbach adds, "The first alert with Cash: he looked at me — but didn't bark. The *look* is what organically started with him. I built in the bark."

McGee suddenly runs off to follow her fast-moving charge. "This way, sweetheart! Did you find something? Good girl." McGee reaches her and looks down. "Oh, spit it out. So," she turns back to me, jogging up to them, "if I am not on top of her, she will *self*-reward. And with black truffles, she will occasionally eat them. So that's what that was." (She has also been known to eat bugs, snails, and the large, thick banana slugs of the forest.)

For the next thirty minutes we all engage in an exercise in chasing after dogs, McGee ready with pieces of hot dog, chicken, and cheese to ply Lolo with before she self-rewards again. They gather a dozen truffles, leaving behind countless pinhead-sized ones that the dogs nosed — but hardly worth keeping. Both dogs hunt naturally, without request, as if once they see the forest they know the game. The dogs alternate deep sniffing and panting, tongues very red, pulsing, and curling. In counterpart, the handlers keep up a stream of encouragement. Lolo tears through the bramble, running low to the ground, her rear legs moving together and giving a little hop to her step. "She's what you might call a *rustic* Lagotto," McGee says, watching her. "She is more compact, heavier, and squatter" than other lines of the breed. It seems not to affect her ability to maneuver — or her sometimes sophisticated search methods. At one site she "brackets" the truffle, going back and forth as though diligently examining each square on a grid until the truffle is found.

Eventually the handlers call their dogs off the search: "Free dog!" This slows them but does not stop them. "Lolo, wait for me please! Lo? Lo? Can you come? Good for you!" Lolo, pausing, comes, then scarfs down the contents of McGee's palm. "Searching behavior with Lagotto in general tends to be pretty hardwired; it's rewarding itself. It's like taking a herding dog and watching cars go by: it's like chasing cars, basically." For the first time, Lolo seems to notice me. She suffers a tickle under her chin. As I lean over I catch a whiff of truffle breath.

Some think the North American black truffle smells of "pineapple and a little bit of chocolate"; others simply call it "absolutely disgusting," with a texture "between moist parmesan cheese and ground almonds." They are not "mushroom" smelling fruits; instead, truffles are variously described as like smelly socks or hot dogs; cucumber or green apple; molasses or musk; garlic or gasoline. What at first glance looks like a rock covered with dirt reveals itself to be a black truffle, its surface marked with warts. Inside it is peppered white, generally hard but with plenty of give under a thumbnail. If it's ping-pong-sized, you may have found fifty bucks; if it's a grapefruit, you've hit the jackpot.

The truffles are the fruit, bearing spores, of a fungus: the fungal "organism" is actually a network of filaments called mycelium that wraps around tree roots. Given the value people place on the fruits, attempts have been made to agriculturalize the process by planting likely host trees and inoculating the roots with spores. Thus far this method is not sufficiently successful or controlled that Lolo has to worry about her job security. (And even on a tree farm, dogs are needed to root out the stock.)

For training dogs, McGee and Rosenbach make a solution from grape-seed oil — relatively odorless among oils — infused with the headspace of the various truffles that they find in the region. They don't steep them in the oil — "because botulism," McGee points out. They just let the oil and the truffles hang out in a closed area together, and the fat soaks up the odor. If they are affiliating for a short time, the oil is relatively odorless for humans, but the dogs have no trouble detecting it. On the other hand, McGee warns me against keeping a truffle unsealed in the refrigerator (essentially a closed box) for any length of time. The volatiles are picked up by any fats — cheese, butter — and "it will truffle everything in there. It can be gross."

We head to another site to let Rosenbach's other dogs have a go. Another fir forest, this one is littered with ferns and salmonberries, proudly poking out like new raspberries. The undergrowth is dense; low dead fir branches catch my hair and backpack. The "open area" that we are headed to has sword ferns that climb up to my neck; in another month this will be impassable.

Da Vinci, a Belgian Tervuren, and Lolo are on the hunt. Da Vinci is a calm dog, with the stately bearing of a larger dog. Though he performed a short protest howl when left in the car, his approach is measured. He is the least experienced of the dogs out today, having searched in forests for only one year. Rosenbach keeps an extra close eye on him. *"Trovarlo,"* she calls out — "find it" to the *tartufaro.* Da Vinci apparently speaks Italian, for he heads straight to a tree and noses around, doing nearly a headstand as he reaches under himself. His style is less frenetic than courtly; he kicks up some dirt, snorts, and waits for Rosenbach. "He's very literal," she says, coming to see his find. As I approach, the whole space smells cloyingly strong: truffle air.

We head in deeper, the two dogs zigzagging through the dappled sunlight. At one point they pass each other, pause to wag and perhaps share information, then continue their respective ways. As we watch the dogs, McGee muses, "A lot of people have the idea that *oh, the dogs just go do it.* It's like, (a) they're machines, and (b) you just train them once and that's it — or that I can train a dog and give it to you, and you'll be able to operate it." While they are called truffle dogs, the handlers are integral to the dogs' success. Both handlers are highly aware of what their dogs are doing in the forest, and work to subtly help them if they encounter an obstacle. They have to continually read the dogs' behavior, a conversation without words. If a dog who lies down as an alert isn't alerting, but is paused, it might be because there is a bed of nettles under his body; if a dog is digging but not finding the truffle, the handlers might move some dirt around themselves to unearth it. And training, as with all the detection dogs, never ends. If a long search ends with no finds, a good handler may manufacture a success, tossing a truffle nearby, to give the dog the sense of fulfillment. "My job is to make sure that they are *always* satisfied" with the outing, Rosenbach says. "He should always think he's a genius, a rock star."

I head Lolo-way. McGee suddenly breaks into a run — "Good girl, good, good for you" — and approaches Lolo at once rapidly and stealthily, careful not to startle her. She bends over Lolo with her spade and pail out. Declined downward, Lolo is only a little curly rump, tail an immobile exclamation point, madly digging into the ground. She snurfles, a half-grunt half-sneeze, and kicks mounds of moist, dark earth toward McGee, now kneeling next to her uttering assurances, "Good girl, want help? You're doing a good job." Lolo digs steadily and heartily, keeping her head down, occasionally pausing to push her nose in as far as it can go, snort-inhaling the smell. The curly hairs of her dangling ears are painted with inky brown earth that has never seen sunlight. Then the slightest of pauses — McGee notices, *"Did you . . . ?"* — and Lolo lifts her head out, up to her eyebrows in dirt, her panting tongue curling up at its end. "Wooooow!" McGee says, looking into the hole she has dug. She keeps Lolo at arm's distance and calls me over. While Lolo is inhaling her cheese reward, I peer into the hole. The fertile, earthy smell of soil and green plant life is suddenly supplanted by the smell of earth from the other side: piercingly fungal, on the edge of sweet and revolting. There are roots and clusters of displaced moss, and McGee points out a dark dirt clod, about an inch and a half across. It looks terrifically undistinguished. But to Lolo's nose and McGee's eyes, it is clearly a handsome sized black truffle. "A nice size," McGee says admiringly, fingering it and plopping it into her pail. "Thank you! Good girl! Yay for you!" Lolo is off running again.

The day is waning but Rosenbach's third dog, Callie, needs a go. Once the door to her crate is opened, she is out like a rocket. She gets a ball in her mouth, squeaks it twice, and looks at Rosenbach expectantly. They disappear into the trees. McGee slips me a small plastic container, sealed. It contains a brown, lumpen mass and a paper towel. I get into my car before daring to peep inside, then stash my precious cargo in a large ziplock bag. Heading back to the city by thousands of acres of forest, it feels that truffles might be *every*where, under each fir that proudly guards the dark forest. An hour later, the scent in the car is overpowering. I open all my windows and spread a trail down the highway for anyone with the nose to follow it.

-PERSON-

They haven't got no noses,
The fallen sons of Eve;
Even the smell of roses
Is not what they supposes;
But more than mind discloses
And more than men believe.

— G. K. Chesterton,
"The Song of Quoodle"

Early in June, I beeline for Central Park, head two miles south, and stop under an American linden tree, *Tilia americana.* I did so on recommendation of Raymond Matts. Matts is a fragrance designer, maker of a line of niche perfumes. While he concocts odor combinations — scents that are never found in a natural setting — he also celebrates odors that appear naturally. Like that of the blossoms of the linden tree in June.

Its branches are beginning to bend under the weight of uncountable tiny, cream-colored flowers. They grow in clusters, each flower really an explosion of petals and dozens of tiny stamens. As I loiter underneath the tree, the odor is heady: honey-sweet and full. A cloud of perfume greeted me well before I saw the tree, but the smell was so widespread that I could not have found its source without knowing where to look.

To stand for ten minutes under a linden is to have one's senses undone. It is to swim in odor, to bear the blows of friendly attacks to the face. It is probably something like what it is like to be a perfumer.

If there is an equivalent for humans of the heady smell of truffles to dogs, it may be perfume. We have been rolling in it, so to speak, for thousands of years. Cuneiform tablets provide evidence that fragrances were added to wines four thousand years ago; ancient Greeks and Romans perfumed their clothes chests and, very specifically, their bodies: "mint . . . for the arms; palm oil for the face and breasts; marjoram extract for the hair and eyebrows; and essence of ivy for the knees and neck."

The popularity of fragrances has waxed

and waned since then, but given the fears, at various times, that bad smells brought disease, perfumes stepped in as a safeguard against malodor. At times perfumery was considered a perfectly reasonable substitute for a cleansing bath. On the other hand, the increase in public spaces in the late seventeenth and early eighteenth centuries — which brought more attention to "personal space" — led to a movement against perfumes. As sensory historian William Tullett has noted, a fragrance fundamentally "extends [one's body] into the space beyond which it should" — and into the personal space of others. Through an act of Parliament, perfume was deemed false advertising in England: if a woman "seduced or betrayed into matrimony" any gentleman by use of scents (or, it should be noted, artificial teeth, false hair, or high heels) the marriage was rendered null and void, and the woman subject to misdemeanor punishment.

Let us acknowledge, here, the strangeness of perfume. A perfume — from the French, *fumigate* or "to smoke thoroughly" — is an odorous chemical combination, usually suspended in alcohol. It is an olfactory intrusion, an odor without a reason, and what is more, it is a kind of "dishonest" signal, in biological terms. That is, animals, earth, flowers in bloom all *smell* — like themselves. We smell the skunky presence of an animal startled or killed, and we expect that it comes from . . . a skunk. Smell indicates presence. By contrast, perfumes are concoctions, odors separated entirely from their origins and mixed together, sometimes hundreds at a time. Instead of smelling the "rear end of an Asian cat," as Luca Turin describes it, when we smell Chanel No. 5, we smell the soft velvet whole, reminiscent to many people of their mothers or grandmothers who wore it habitually. But it includes the greasy rear-end secretion that the animal called civet* uses socially as a fear response or to mark a territory. The nocturnal, low-slung civet, which looks like the unlikely love child of a feral cat, mongoose, and raccoon, was formerly kept captive in small cages and bothered in order that it might produce this secretion; today, synthetic compounds have largely replaced this practice.†

* Not a cat, but one of various species of the *Viverridae* family, which also includes other animals you have not heard of, such as the genet and fossa.

† Chanel No. 5 was also one of the first fragrances to use synthetic components — aldehydes C10–

Those who work with fragrance — as perfumers, mixing the odorants, or designers — and those who work with wine — as sommeliers or tasters — are widely considered to be the best smellers around. Their powers of detection, distinction, identification, and simple *noticing* are used professionally, and inevitably change their way of experiencing the quotidian odors that they share with us novice noses. We smell a rose as a rose, but Coco Chanel claimed to smell "the hands that picked" the flower as well.

There are myriad routes to becoming a perfumer, but if you want to work at a big house, evaluating or making perfumes, there are internal perfumery schools, with a curriculum that involves learning to distinguish blindly between hundreds of raw materials. Ron Winnegrad, the director of the perfumery school at International Flavors & Fragrances (IFF), designed a multiyear curriculum for the five or six students he has at one time — who "show promise on olfactory and psychological [application] tests." Winnegrad tests the students every day on raw materials. They are also tested on "duplications" — they must try to duplicate

---

C12, imparting "freshness" and "citrus pith" — when introduced in 1921.

a fragrance that is on the market — and, if Winnegrad adds or decreases a little of one ingredient, the students are tasked with identifying what has changed.

Not only are most perfumers not super-smellers, many have specific anosmias: that is, they are entirely unable to smell some odors, such as musk. But "although they can't smell it in isolation, they can tell if it's been added to a perfume," Leslie Vosshall tells me. "They can paint with it." And they pay attention to smells, bothering to try to remember what the building blocks smell like: flash cards where one side is a puff of phenylethyl alcohol (a floral odorant), not high school vocabulary words. As with other forms of expertise, the perfumer's brain reorganizes itself with practice; indeed, one could say that it is the reorganization that makes the expert. While a part of the brain called the primary olfactory — or piriform — cortex is active in all people when smelling, fMRI images show that professional perfumers' brains process smell differently than others. They seem to rely less on memory areas; instead, perceiving or even imagining odors is a more automatic process for expert perfumers.

As important, expert perfumers helpfully remove evaluation of an odor from their consideration of it. It is not that they do not themselves experience smelly socks as "bad" and coffee as "good"; it is that they are more interested in whether the smell will work well in admixture with other smells to make a perfume. "I *hate* the smell of coffee," Raymond Matts says, frowning — by which he means not only that he finds it disagreeable on its own, but that it interferes with good smelling. On the other hand, many perfumes that he no doubt admires have evocations of coffee deep in their olfactory heart. To say nothing of the civet, ambergris (formed in the intestines of the sperm whale), and castoreum (beaver perineal secretion) of the fragrance world. "Has the highly distinctive pungent odor of sperm whale feces," a researcher of the whales writes of fresh ambergris. The very common "musk," used by male deer to mark territory, is now accepted by most people as the clean smell of "laundry": it is used as a maskant to cover up what Matts describes as "the horrible base" — enzymes needed to clean clothes — "the smell of eyeballs; the worst fish you can imagine." (Use only "unscented" products? Until recently "unscented," too, was a scent — used to cover up the musk.) Clinique's perfume Happy, which Matts worked on, is based around lily of the valley, but there are also what he merrily describes as "pukey" notes — "that keep the fresh floral notes trailing in the air," he explains.

For a "nose," as fragrance experts are sometimes called, Matts has an ordinary looking smeller: slightly broad, classically nose-shaped. It is partially hidden by the thickness of the bridge of his eyeglasses — resting, it should be noted, just above where the epithelial tissue with his well-trained receptor cells lies. Matts is youthful in appearance, belying the thirty years he has worked in fragrance. His career has ranged from a job as evaluator of fragrances of products like laundry detergents and industrial soaps, to working at various houses and helping to design Liz Taylor's White Diamonds perfume and T, a fragrance of Tommy Hilfiger. T was one of his remarkable hits, though Matts says that it might be the least respected one. He describes smelling the fragrance on a man seated next to him, and asking him after it. "The guy denied it," Matts says. "But no one [but me] ever used saffron to that level." The nonperfumer can be forgiven for the blank expression she might wear at that comment.

Matts has no facial hair. It does not become someone who deals in fragrances. While Michel de Montaigne noted that his "mustachios, which are full" served admirably as vehicles to keep wanted and unwanted smells resting under his nose for hours, the perfumer cannot afford this distraction.

When Matts introduces students to smelling, he begins with "families" of smells, combinations of "accords" — from three to ten odorants that combine well. At his class The Technique and Language of Perfumery at the Pratt Institute in New York, an enticing olfactorium — a kind of traveling trunk that houses forty-eight bottles, each in its own soft housing — sits open on the table. Each bottle holds an odorant, one of the palate of colors used to make a fragrance in the "Oriental" (vanilla, sandalwood, patchouli) or "Fougère" (lavender, vetiver grass, oakmoss, florals, coumarin) families, for instance.* A blotter, a long strip of paper with which to gather scent, is bent and dipped into the bottle. Matts dips his in a

* You might know an Oriental: Johnson & Johnson's baby powder. The fact that this is more familiar to most readers, I'd hazard, than the description of the family of odors, is a testament to how our vocabulary of smells has been hijacked by those used in retail products. It would be as if all our musical knowledge were based on commercial jingles.

bottle with the number 16 on it. He pauses significantly to let the paper dry — otherwise "you're only smelling the top notes," the scents that volatilize and waft off first. Perfumes are made to have three tiers of "notes," usually called top, middle (or heart), and base (or background). The top is what you smell when spritzed as you race through the perfume section of the department store; the base is the odor that is on your clothes the next day. My mother's now-vintage Écusson eau de cologne will not smell the same as when she bought it: it oxidized every time she opened the bottle. The top notes — citrus, aldehydes — break down, and the heavy, base notes of oakmoss and vanilla might stick around. "It will *resemble* the fragrance it was," Matts says.

"Mmm," he says, after drawing the number 16 blotter under each nostril. "Cedarwood. From South Morocco and Virginia." This is one of the "woody" notes. Like a tea expert who can identify not just the origin of the tea, but the time of year it was harvested and whether it was planted near a plum tree, a fragrance expert detects real differences in the quality and source of what to the rest of us appears to be one category of smell. This one is "pencil shavings" or "hamster cage" — less smooth than the air that rushes out on opening your grandmother's cedar chest. "Can you smell the waxiness?" He lifts the blotter to his nose, sniffs, pulls it away, considers, lifts it again. "Think about a Crayola. Do you see the dryness and depth?" Even over a few minutes, the sharpened-pencil smell resolves a little, gets smoother.

With another wood note, Matts says confidently, "You can smell the grain of the wood" — as though it has a distinct olfactory fingerprint. Sandalwoods come creamy or raw. Some woods are oily: the synthetic Norlimbanol, dry, harsh, and strident. Dried patchouli smells of earth and roots. Another smells like nothing more than fresh-cut twig to me. But to Matts: "Velvet. On top, it's the transformer of the electric train set when you turn it on — the metallic odor."

Reflexively, one might think that "wood" smells, in essence, of trees. No longer: the range of woodiness suddenly appears vast. "I'm always trying to get the smell of the construction site: the circular saw going through wet wood," Matts confesses. "I haven't got it yet."

Matts is frequently trying to capture a found smell to bottle it. When he had a young baby, he became entranced at his smell: milk coming through skin. He thought the ubiquitous baby blanket covered in spit-up "smelled good" — the sour notes evaporated, and leaving the creamy essence. He spent some time trying to capture the smell, and gave up after his wife balked when he brought out a hypodermic needle.

Indeed, there is a way to capture the kind of essence that enchants so many of us: the odor that emerges from the baby's head, from the tomato vine, from the dried and raked leaves. Perfumery could be described as initiating with that urge — although it has gone far beyond simply trying to capture natural smells. Various methods have been used over the years, from enfleurage (similar to the truffler's method, absorbing scent of, say, a rose into a fat), extraction (rinsing the source with a solvent and extracting the oil), distillation (creating a vapor by heating and then chilling to separate the oil), expression (crushing citrus peel, rinsing, and cold-pressing it), to infusion (macerating a plant in alcohol).

Better for the non-maceratable odors (like the baby's head), though, is headspace technology: a glass bulb is put over the item, capturing the air that wafts naturally off of it onto an odor-absorbing material. The captured odor can then be analyzed in a lab with gas chromatography, and a perfumer can try to re-create it. The possibilities of this method are most fantastically represented by Roman Kaiser, fragrance and flavor chemist for Givaudan, for instance, who traveled the world for a decade to preserve the headspace of endangered flowers, should the source disappear. Self-described "professional in-betweener" Sissel Tolaas has amassed a library of thousands of odors via headspace, including the smell of her daughter. Matts describes using the technology to capture the distinctive odors of Cool Whip and Wheat Thins when designing a fragrance for Abercrombie & Fitch.

Standing in front of an array of tiny bottles, Matts picks one up and spins off its black cap, bearing the number 42. It's a leather note, and begins sharp, almost medicinal. Matts smells the blotter and a faraway look appears in his eyes: a smeller's look. He is seeing something. He begins speaking, describing the process of creating the odorant: "Originally it was made by infusing scraps of leather — which itself smelled of the tanning, done with birch-smoking." On saying *birch* the woodiness of the odor rings brightly, like a child called on in class. "Think of August in Amish country," Matts says. "Tobacco leaves drying, hanging by barn slats. That juicy, sweet smell in the air." With his words, the other odors in the class jump to attention, one by one. Dried tobacco leaps forward, the moistness of the end of a well-chomped cigar; the aura of barn, fecal and weathered. Indeed, it has been suggested that part of the practice that transforms a budding perfumer or wine taster into an expert is the need to "verbalize their olfactory experience" — to put a language to their experience. Both wine and perfume have their own jargons — for perfume, the *Oriental*s and *Fougère*s — but all individuals in the field also generate their own vocabulary to remember and classify scents, serving as imaginative touchstones. These landmarks are a significant part of not only experts' discrimination and identification abilities, but also their ability to invoke a smell without the smell being present.

Matts's linguistic landscape for smells is rich with evocations of familiar objects and childhood memories. Ylang-ylang, a floral note, is "Necco wafer"; jasmine, "the horses of Central Park on a hot day." Over a few hours, he references rubber doll heads, the smell of crayons, white paste, basements, latex gloves, a Lionel train set, wet paper, carrots. His specificity is humbling: to Matts, a floral note doesn't smell like the flower, but "like the air when you're walking by the flower."

As language can be used to bootstrap smell sensation into memory, the other senses are also useful. Give a roomful of people a bottle with a note of heliotrope, a purple summer flower, and rather than asking them to describe the smell, ask them to give the smell a shape and a color. Something surprising happens: almost everyone smells it as yellow-pink, or a variation thereof, and sees it as round or almond-shaped. The aldehyde C12, used to impart a sense of "freshness," is most often experienced as some form of blue and square. We share an unspoken, untapped knowledge of smell associations that we are simply rarely asked to relate.

After a few hours of smelling, the room we are in becomes a monster of odors. An overwhelming wash of *something* disorients my nose, like I need a psychic sneeze. I begin to get a little wobbly. Olfaction is a kind of "change detection" system, so if nothing changes, it turns itself off. But if new smells keep getting distributed around you — even variations on an old smell —

the nose does not adapt, and desperately tries to keep up. With every bottle we open, we are restimulating those receptor cells trying to catch a few winks.

Before I leave, Matts makes a recommendation for how I might become a better smeller. "This summer," he suggests, "pick things up. And just squeeze them between your fingers" — a kind of on-the-fly extraction of odor — "and smell."

I decide to do Matts one better. Not only do I begin bringing things to my nose, I start bringing my nose to things: I again bend down to the sidewalk; I sidle up to trees and then nose them. I follow the advice of an animal-tracking handbook to "calibrate" my sense of smell by sniffing at different walking paces, in different directions, and at different latitudes (nose-height and toes-height). "Go for a very slow walk," the authors suggest. "Stop every ten steps, or every time you get a new rush of scent that seems different. Do this for about ten minutes," alternating regular sniffing and doglike sniffing. I do.

To counter, or perhaps encourage, these habits, one day my husband delivers to me a large package.

The box is designed to appeal to the buyer

of Wine with Fancy Labels: it is dense and solid. LE NEZ DU VIN reads the cover. An outer case, lined in fire-engine-red bound cloth, slips off either side to reveal another case, redoubling the mystery. The inner case splays like a book; inside are fifty-four small, rectangular bottles, each snugly secured in its own fitted slot and beguilingly filled with mysterious liquids. Some are perfectly clear; others, amber, or, worryingly, lake-dark.

A box of bottled wine notes is the kind of thing that must be approached slowly. The first day I sit down with my bottles of curious smells and rub my hands together, a rare gesture of real excitement: *oh, the smells that await me!* But then I stop. I am already smelling, and it is a cloyingly sweet, strong smell. And it is coming from me. My shirt: it has just come out of the dryer. A major blunder; now I am steeped in a musky lavender, smelling of "clean laundry." I strip, find a shirt that I'd aired out by wearing the day before, and begin anew.

I open the first bottle and bring it bravely to my nose. Left nostril, sniff; right, sniff. *Lemon* pops up in my head at once. I check my experience against a deck of scent note cards and move to the next bottle, feeling self-congratulatory. Bottle number two is also of the citrus family; I'm going to say *orange.* Wrong: grapefruit. I return for a second sniff. Okay, I can smell it now, with an image of a grapefruit in my head, imagining its heft in my hand and the thickness of the skin yielding under my thumbnail.

I continue through the deck, joined periodically by a member of my family or a dog, both intent on sniffing, too. The dog, Finnegan, waits patiently for a bottle and then, when I offer it to him, sets to his inspection like a professional. Without reading too much into it, I think it is fair to say that he looks genuinely puzzled. He shoots me a look, then sniffs again, ending with an exhale that makes the bottle toot like a one-note recorder.

My early success with "lemon" is not often matched through the run of the bottles. I mistake "toast" for "vanilla" — impossible, I would think. I confuse "peach" with "melon"; every form of berry (and there are six) smells to me just like "candy." "Saffron" and "hawthorn" ring exactly no bells in my sensory memory. While "smoke" is a cinch, I struggle with "leather" and "butter." I have the classic tip-of-the-nose problem: I *know this smell,* but I cannot name it.

For professional smellers, naming the odor is not the goal, per se; it is simply a step toward more understanding. But for my novice nose, a name is necessary. Over time, I invent a strategy to try to come up with the name: an image-matching complement to the memory-search I learned when participating in the smell study. Recognizing that context matters for identifying scent — it is much easier to identify the smell of "apple" in an orchard, rather than in a spa — I decide to use my visual imagination for good. As I sniff, I close my eyes and conjure random objects to appear in my mind's eye. They begin to appear out of nowhere, hallucinations out of my conscious control — a ceramic sink, a wool car coat with a sheen of frost on its shoulders, a school classroom pencil sharpener — and when one looks promising I zoom in on it. A small tree, bare of leaves. I hold it tight and check if it matches the odor rollicking up my nose. Not quite. Switch it out for another tree, hold it close, sniff . . . repeat.

In this way, I make my way round to "pine." An image of the prototypical pine tree shimmers into my head; the odor of note number 35 suits it. After I identify a scent, it's *obvious,* like a dumb joke badly done.

And so begins my practice, a kind of smelling introspection, refined over the months. I like to start my day with coffee, but on the recommendation of Ray Matts, I have to give up this habit, or plan a session later in the day when I will be relatively coffee-free. Ideally I would rise, have a glass of water, and sit to smell. But often my stomach grumbles, or the dogs lick me to go for a walk, or my son is raring to start the day, and I have to put off the smelling practice to another hour. This is a tricky business, because, as I learn, the day is actually full of smells near my face.

My hair is redolent of the perfumed shampoos I wash it with, as the fragrance gets caught between the layers of hair and the oils grip the shafts (for this reason Matts recommends putting perfume in your hair over dabbing it on the wrist). My face has a vaguely Hawaiian suntan-oil trace from my preferred moisturizer. Hand soap smells hugely — and every time I lift my hand to my nose I get a whiff of whatever it was. Soap in public restrooms begins to worry me: what impossible stink will I be painting myself with for the mere interest in cleansing my hands of bathroom germs? Anything in my mouth is off-limits. Since taste is mostly smell, any echo of the rosemary bread or cucumber or peanuts that I consumed an hour ago presents a major roadblock to smelling the wee bit of odor wafting out of a bottle.

There are plenty of days I simply cannot spend an hour with The Smelly Box. A stuffy nose is a disqualifier; relatedly, simply being chilly makes it harder for me to concentrate on any smell at all. Even when I am warm and clearheaded, the episodes are brief. Our noses tire easily and shut down with too much stimulation. I hoped that I could conquer at least the nose-fatigue with training.

Once I have made it all the way through the deck of bottles, I can no longer be surprised at the odors, so I practice distinguishing between smell groupings: animal, vegetal, toasty. I sit down with three or six bottles in an odor group, shuffle them around, and try to identify each blindly. When I get nose-weary, I take a cue from Matts: I smell my own shirtsleeve, at the crook of my arm. This olfactory interruption with, presumably, one's own odor works as blinking does to restore vision after staring at a bright light.

To resolve the ongoing orange/grapefruit confusion, I step into my kitchen, where I have both grapefruit and oranges. I hold each to my face. To my surprise, the actual grapefruit smells "citrusier" than the orange, which I think of as the model citrus. I have the same difficulty with two of the stone fruits voted most likely to appear in my mouth: peach and apricot. While they are dramatically different fruits in my mouth, I am unable to reliably smell the difference.

It is at this point that I visit, profoundly and fully, my lifetime smell agnosticism. These are among the simplest of smells. They are common. I have ingested thousands of oranges and grapefruit, peaches and apricots, exactly for the reason that I relish their piquant smells and tastes. Was I not really attending even to these most loudmouthed and big-breasted of smells?

Nope. I have, apparently, spent my "eating" life simultaneously training myself not to pay attention to the subtle scent differences between foods. Each orange I encountered was, well, orange-tasting, and I did not have to check to be sure I was not eating a grapefruit — and vice versa. I could tell the difference in my mouth; indeed, I enjoyed each one differently. But I had gotten lazy about noticing anything but that it was a confirmed "citrus fruit." So my citruses have collapsed into each other. I untrained myself out of what children notice naturally (try to sub a grapefruit for an orange and see how a child reacts). I grew out of noticing.

In this, I have great company among English speakers. While "lemon" had, in my nose, preserved its own distinctive spot, in cross-cultural research asking people to identify common smells, English-speaking subjects were likely to identify an out-of-context lemon odor as "air freshener," "berry," "Magic Marker," "citrus," "hard candy," "some kind of fruit," and, the most depressing of all, "lemon-fresh Pledge." For some people, lemon-fresh Pledge is a more vivid olfactory memory than an actual, honest-to-God lemon.

These subjects' vague characterizations of smells contrast with the whiplike precision, relatively, of the olfactorily-sensitive Jahai people of the Malay Peninsula. Though probably less fluent in the odors of lemons, turpentine, onion, cinnamon, and the other odorous stimuli the researchers presented them with than the English-speaking subjects, they were much better at naming the smells.

It takes me days to get the citruses reliably right. Now I can pull it out like a great party trick — though I have no doubt plenty of people would have had the same ease on the first go. The ability to detect the presence of an odor varies widely from person to person; we are similarly individual in our ability to identify the odor. But while the former has to do with our genome, the latter has to do with our life. I am much better, from the get-go, at the smells I have registered as dislikes: almond, smoke, licorice. My attention is, it seems, well-tuned to spot the unlovely.

The Le Nez du Vin box itself, with all the bottled scents cloistered inside, is a hugely busy odor. It is the wine that you would never sip, and evidence that pleasant scents are not additive. Put together, the sum diminishes the parts. Walking away after a morning episode comparing almond, walnut, peach, apricot, cherry, and prune, I find that the whole room smells like prune. I step outside with Finnegan; a wind wrests the screen door from my grip. His nose rises to attention at the passing air. I smell . . . prune. I have been afflicted with a prune noseworm. The smell sticks with me for an hour. As I continue to challenge my nose, eventually sorting nine at a time — *cedar, rose, green pepper, thyme, vanilla, cut hay, acacia, caramel, banana!* — and finally all fifty-four, I seem to become more susceptible to the occasional noseworm or errant olfactory hallucination. I do not worry: they seem more to mark my increasing sensitivity to odor than any disorder. Chronic hallucinations, or phantosmias, though, are far less gentle. They are frequently reported by psychiatric patients or people who have experienced head trauma, and less often take the form of lovely prunelike odors than of burning, foul, spoiled, vomitous, or rotten smells.

Within a few months I have made a library of smell notes in my head, with idiosyncratic descriptions that reliably help me along: prune is vanilla, dried out; apricot is more cloyingly sweet than peach; cherry is cough syrup; walnut is dry almond; "lees" (the yeasty sediment left at the bottom of a wine barrel) smell like "wine"; violet is small hard candy, dusted with sugar, in an oval tin with a decorative top; bell pepper is "green." I have become a Nez du Vin expert. What I am not, is any more sensitive to the notes of actual wine. Each glass I bring to my nose I sniff, I swirl, I nose almost to the point of touching the liquid. I do not smell apricot, peach, grapefruit, orange, jam jar, mushroom cap, or Gauloises. I smell — not soda pop, not water. Definitely wine.

But that is about it. And so I fly across the country to meet John Buechsenstein.

Buechsenstein, longtime winemaker and wine judge, stands at the head of his class at the Culinary Institute of America in St. Helena, California. Drought conditions have rendered Napa Valley preternaturally yellow: hills covered with a fuzz of browned grass roll into soldiered rows of trellised, green-leafed *Vitis vinifera* grapevines. Inside the so-called CIA, a huge bricked structure that was formerly a wine vault, the classroom is deeply cool: a wine refrigerator in which students now chill. The tiered lecture hall tables and chairs face a lectern and whiteboard in a familiar collegial way — but each table is lined with spittoons and sinks, and eight perfectly poured glasses of wine sit at its center. This is the Sensory Analysis of Wine class, boys and girls, and you will be spitting.

The room exudes spilled wine, the intermixed aromas of the first samples, ranging from light yellow ("pale straw," to Buechsenstein's eyes) to darker ("rose," "garnet," and the "purple" typical of a petite Syrah). There is plenty to *see* in wine: from the relevance of a brilliant or slightly hazy pour, to the faults (sometimes rendering the wine undrinkable) indicated by a cloudy glass or one with a constellation of crystals at its bottom. And of course there is plenty to taste. But "smelling is the most glorious part of wine tasting," Buechsenstein says, smiling. Buechsenstein smiles a lot, furrowing his brow and nose while he does, as though working to keep his glasses up. And he smells a lot. The alacrity with which he will stick his nose into a glass is at first surprising, and then contagious. His mandate in this class, which he developed and has taught for three decades at UC Davis, seems to be to give wine enthusiasts and professionals the means and mettle to stick their noses into glasses and really sniff.

For Buechsenstein, this begins with simple attention. "If you do nothing more than pay attention to what's in that glass, you'll be miles ahead" of the person next to you — or the person you were the moment before. To focus that attention, he gets his students — of which, for two eight-hour episodes, I am one — to bring a lot of glasses to their nebs, and to flirt with talking about what they notice. Initially, this is difficult for everyone. Wine smells like *wine,* it seems. We are so used to greeting a glassful of wine by glugging it, we rarely pause to reflect on that moment before it pours past our lips. And we haven't learned a language to describe what we smell.

But of course there is a lot of wine language. "Low-keyed, solidly structured, with a youthful buoyancy" — what Buechsenstein calls the "poetic school of prismatic lumination" — dominates wine reviews and prose. This way of talking about wine seems intended to exclude all but the cognoscenti. Buechsenstein's own descriptions of aromas are a bit more accessible, invoking shared experiences, memories, and conjurings:

"That tasted to me like I just bit my Bic pen open."

"Ooh — St. Joseph chewable fake-orange aspirin! I take one of those every night."

"This is the charred log from your campfire, smelled the next morning."

. . . and familiar food smells:

"chopped celery"

"when the pot boils over and beans burn

on the stove"
"baked bananas"

. . . and just plain effusion:

"Now that's aromatic to beat the band."
"Whoa, fruit-and-vegetal double whammy."
"I put it up to my nose and it was like Dorothy-landing-in-Oz-Technicolor."

To enable his students to get to Oz, Buechsenstein concocts a series of "flights" of wine — usually six or eight glasses to be compared and contrasted. Many consist of a neutral wine and variations therefrom, doctored with flavors that one might find in the wine. A red might be steeped with a sliver of bell pepper, another with a couple of peppercorns, a third with some crushed raisins. To pull out the notes of a neutral white, he spikes a glass with a bit of canned asparagus juice (just as horrifying as it sounds) or marinates it with a martini olive. On another occasion Buechsenstein cuts a white wine with varying levels of sugar or acid, and a red with acid and enough tannins to seal my teeth to the inside of my lips. Over a few days I come to pick up on — if not always correctly identify — "all that funkadelia" that Buechsenstein finds in wines.

Watching him approach a glass of unknown wine is its own master class in tasting. He eyeballs the glass and waves his nose over its top with a perfunctory sniff. Picking up the tulip-shaped glass by the stem so as to not inadvertently warm the contents with his hands, he tilts it to the point of near-dumping to see its color and makeup. Then an assertive swirl of the wine, still gripping the stem. The swirl washes high on the sides of the bulb, creating petals of pink and violet and red that replace themselves at once. In the same gesture he places the glass directly under his nose — or, more precisely, places his nose directly into the glass. "Five bunny sniffs," he says, audibly sniffing, then hoovering a last sniff. His gaze is directed to an invisible point two feet south and in front of his head. Swirl-sniff-repeat. Then, suddenly, he brings the lip to his mouth and takes a sip. We have all sipped drinks — but never with the complete confidence that we will not spill anything on our shirt that John Buechsenstein has. This wine does not go the way of all wine that enters my mouth: down the gullet. It is audibly swirled around his mouth. Though his mouth is closed, we can almost see it going over his tongue, across the hard palate, around his teeth. It is gargled, to its great surprise. He pulls his top lip down as though clearing his nose. His moustache quivers. And then he unceremoniously — but actually ceremoniously — turns and spits it out.

This tasting technique has been refined over the dozens of years as winemaker and as graduate student at UC Davis, which has a renowned viticulture and enology program. As a TA at Davis, the methods for learning smells were sometimes a bit ad hoc: "When we had undergrads who couldn't smell Brett" — *Brettanomyces,* a yeast that leads to a classic wine fault — "we'd go to the Davis polo team and we'd say, *Give me your funkiest, sweatiest, ready-to-throw-out blanket,*" Buechsenstein describes, rubbing his hands together. "And we'd stuff it in a ziplock bag and seal it up. And say, *Who was it who doesn't know that smell?*"

More formally, smellers now consult the wine aroma wheel introduced by Ann C. Noble and her colleagues in the 1980s. In the world of wines, it is the primary color group: the reds, blues, and yellows that fundamentally characterize all wine odors. There are twelve main groups — fruity, floral, pungent, earthy, chemical, vegetative, nutty, caramel, woody, oxidized, microbiological, and spicy — with secondary and eighty-six tertiary subcategories beneath them. If you're smelling "chemical" it could be plastic or tar, wet wool or garlic, rubber or mercaptan, the odor added to natural gas. "Fruity" covers everything from lemon to strawberry jam. In the academic paper introducing the wheel, published in *American Journal of Enology and Viticulture,* the authors suggest other ways to reproduce the aromas in wines for beginners. Your butterscotch Life Savers will be useful for generating the caramel aroma that might come up in a muscat. Wet wool added to a neutral white will perfectly mimic that "wet dog" smell of corked wine. Have any extra roofing tar? Let it sit overnight in a cup of wine; that will give you the sense of tar that a Rhône-style wine might exude. Only seven or eight granules of powdered tropical punch Kool-Aid get you that fruity smell in some reds. A single 5×10 mm piece of bike tire, a burnt match, a mildewed cloth? Add 150 ml wine, stir, and taste. Your pantry, supply cabinet, and wood shop are equally useful in this exercise. Elsewhere, Noble suggests putting a single Froot Loops loop or a Handi Wipe (preferably before cleaning the roofing tar off your hands) into a wineglass to get a sense of the aroma of Riesling.

Winemakers employ in-house odor and taste evaluators who not only ensure that a wine is without flaws — elements that some drinkers may like, but that mask the true flavors of the wine — but that there is consistency between yields. They sit in positively pressurized rooms that resist the encroaching of environmental odors and use a descriptive analysis (such as the wine aroma wheel) to characterize the odors in a quantifiable way. Ideally, each panelist's graph of a wine's aromas will superimpose perfectly over each of the others'. If not, winemakers can adjust their wine's aromatics through changing the temperature of fermentation, the length of fermentation, and the time the wine sits on lees. "Flavoring" is never added to wine; a wine's character is simply the result of the grapes and the environment in which the grapes are crushed, then ferment and loiter while awaiting bottling.

Bottling itself is a tumultuous process. "The wine is not really happy about it," Buechsenstein says. Plenty of faults can be introduced in bottling, too — many of those make it to the store and your kitchen. All of the faults can be smelled. If you open a bottle and are greeted with a kind of moldy, musty, old-wet-newspaper, the wine is *corked:* a compound called TCA — more formally, *2,4,6-trichloroanisole* — has developed from fungi or bacteria on the cork and has overwhelmed the wine.*

Tasting is meditation with spitting. Occasional mutterings of pleasure or alarm interrupt the gentle chorus of glasses clinking on tabletops as they are picked up and returned. On a blind tasting I consult my wine aroma wheel in order to induce images into my head to match the waftings from the glass: Is that a really ripe banana? *No.* Something artichoke, green-beany? *No.* A field of hay, laced with honey? *Ooh, maybe!* I scribble a note and clink my glass with the sopranos. Looking up, I see Buechsenstein poking one side of his nose into a glass, as though examining it with a nose monocle. On his glasses hand he has drawn a carbon group in explanation of an earlier aroma; a C=H double bond flashes us on his second finger and pinky for the remainder of the day. He lifts his head: "Did you

* Recent research has shown that, in fact, it is not that TCA causes new, "mold" olfactory receptors to fire; rather, it stops other olfactory receptors — that notice wine fragrance — from firing: the DEET of the wine world.

get it? Think Orville Redenbacher popcorn."

Pop! There it is: a butter smell, diacetyl, one of the side results of malolactic fermentation, a common process with reds. And, since this wine is a straw color — with the honeyed hay — with chardonnay.

On another I get a trigeminal nerve jangle with one wine, the sensation of a match having been struck. My throat is coated. A match is a good stand-in for some of the wine aromas, Buechsenstein says. When it is first struck, there is sulfur; then one gets a smoky odor; finally, a charred mercaptan smell.

Over two days we run through a half dozen training flights. Whites feel easier — their odors are more distinct; in reds, the odors combine and hide in each other. But I begin to find my own language. I get vinyl in one glass; pencil shavings in another. Some of the familiar descriptors — cat pee in a sauvignon; deeply ripe berry in a zinfandel; oaky, buttery chardonnay — become comfortably familiar.

After sixteen hours, my nose has had enough. My mouth tastes only of the tannins of an unpeeled grape. I drive away, board a plane, and look forward to the smells of home and my dogs.

11: Nosed Out

One January day, I took only one of my dogs out. Finnegan. He is the nosey one, the snuffler, the one whose snout pokes deep in pockets and postal packages. Finn was delighted at this rare single-dog outing. He positively pranced over to the car and hopped in. As we merged into rush-hour traffic leaving the city, his mood turned somber. I had no way of telling him where we were going — to what would become his Favorite Place on Earth.

The truffle, disease-detection, scat-detection, and other working dogs serve as evidence to the power of the nose of a trained dog. What about the nose of the one lying at your feet right now, though — that of the well-loved, well-fed, well-pet companion dog? Sure, Finnegan can sniff out a fallen crumb encountered on our walk, and quickly distinguishes between people we meet who-have-ever-had-dog-treats-in-their-pockets and people who have not. But how would he do if faced with a real test of his nose, I wondered.

My lab and I set up such a test for dogs. We started with a simple one: a quantity test. Any owner knows (and research has confirmed) that dogs have no trouble distinguishing one piece of hot dog from five pieces: if given the choice between plates with each, they choose to eat five pieces. It is the rare dog who goes for the plate with one and wags away satisfied. Dogs have perfectly good vision, but certainly any creature with a nose such as theirs should be able to do just as well or better if the plates are covered, and they can use only their sense of smell to discern the contents.

Sixty-four hungry dogs and their people agreed to an afternoon of sniffing and eating hot dogs. Each dog showed great interest in the (covered) hot-dog plates that the experimenter presented to him to examine. Then she put the plates down and the dog was free to approach the plate of his choice. Based on smell alone, the dogs did something I did not expect: they chose randomly. Though they had sniffed the five-hot-dog plate more, they did not use their noses to find that plate again.

Dogs are losing their noses. Other research has supported the odd and disturbing result that companion dogs are not only not using their smelling abilities to their capacity; they are forgetting how to be sniffers. In a human-defined, visual world, it seems that it does not pay to notice all the smells around the house, to sort their way through the world by smell. Instead, the typical owned dog gets a mound of food in a bowl once or twice a day whether he sniffs it out or not. He may be discouraged from sniffing the sidewalk, the lamppost, and even other dogs' rear ends as his owner walks with him — out of the person's disinterest, press for time, or horror. We talk to the dog in words and point at him with hands, but rarely give him smells to learn and live by. The sad result has been that pet dogs are letting their noses go dormant.

But this is easy to change. I look at Finn, idly waiting for me to lead him through his walk, and I decide to change it.

The Port Chester Obedience Training Club building is tucked nearly under an interstate highway, a collector of noise and truck debris. Of more interest to Finn and the other dogs converging on its glass door, though, are the local highlights: a well-visited pile of frozen, blackened snow, graffitied with yellow letters drawn by barely literate urinators; a trail of crumbs that might have dribbled from a dog owner's pocket.

Up a set of stairs marked with helpful yellow pawprints is George Berger, our coach and the current president of the Port Chester club. He greets us cordially, the faintest of smiles wrinkling his bearded cheeks. He wears a fleecy vest, khakis, and sensible shoes. Behind him is a large, bright room, mostly empty but for a smattering of tables, chairs, and agility-course obstacles pressed against three of the walls. The fourth is lined with crates, five of them occupied by dogs. Their owners sit scattered nearby, chatting and peeking over at their charges.

"The game, of course, is nosework," Berger begins. "It's about using the dog's greatest ability, scenting": what the dog trainer and writer Turid Rugaas has called "channel nose."

We have come to this bright, empty room for Finn to smell with impunity. The sport of "nosework" makes a game out of what to working dogs is their occupation: simply using their noses to find something hidden — the "hides." The dog has to improvise, to problem-solve, to correct and redirect himself. Finn has been encouraged and indulged in using his nose — we have spent long minutes at tree trunks and building corners, and I have probably had every inch of my face mapped by his examining snout — which is at least an advantage over dogs who have been yanked from every good-looking smell they nose out. This means Finn's nose is open, ready; but it is not yet tuned. Nosework is the tuning. He will learn that if he uses his nose, and his nose alone (not following what he sees, or what I direct him to do), he will discover hidden treasures and get great rewards (read: copious treats of salmon or cheese).

Berger gently instructs the owners present in the mantras of what is called nosework but is clearly nosegame. First: the dog is the leader. While pet dogs are typically expected to look to their owners for advice on what to do and when to do it (or when to do nothing), here the owners are expected to follow the dogs. "I want them to learn that whatever I or you are doing is BS," Berger says. Eyebrows silently raise a quarter-inch. But, dear eyebrows: to do nosework, the dog has to figure out where the hides are on his own.

Second, the dog must be motivated. At the beginning, this means finding the food treat that the dog loves more than anything — for which he will drop whatever is in his mouth or on his mind and lunge. Later, the reward can be a favorite toy — the one he always yearns to grab — or a game of tug, as with detection dogs. But in all events the dog has to know that something incredibly terrific happens when he uses his nose.

Third, when training, the dog has to find the hide every single time. Until the dogs are experienced at the game, the owners must help them be successful, setting up situations to enable them to find the hide on their own — lest the dogs learn that once in a while, they can give up, look at their owners pleadingly, and get fed anyway.

These are exactly the principles of detection-dog training: independence, motivation, success. Indeed, nosework is modeled on detection-dog training. For the ambitious owner, nosework is even reified as a trial sport with various levels of competition. Just as a narcotics-detection dog might search containers, early-stage nosework dogs learn to find particular odors secreted among a lineup of a couple dozen closed cardboard boxes. As explosives dogs might examine traffic at national borders or airports, elite nosework teams search for odor in outdoor arenas and around cars.

The thrill of this sport, or of watching detection dogs, is that the dog is shedding some of the raiments of domestication. We are reminding the dog — or reminding ourselves — that the dog comes from a hunter who was intrinsically motivated and autonomous. Domestication has allowed dogs to be good sitters-on-couches, but has robbed them of the pursuit of some of their instinctual drives.

What nosework does *not* train is the dog's nose to smell. The nose is perfectly good as it is. It just needs waking up.

Within our small group of six at Port Chester, the constraints that owners have unwittingly put on their dogs' abilities to follow their noses are immediately clear. One dog stands immobile if his owner is still, a perfect *I have absolutely no ideas* look on the dog's jowly face. He is not driven to do anything at all without his owner's direction. In some contexts, this might look like a "polite" dog. In this context, it is heartbreaking: the dog has no internal motivation, no striving to pursue, no ambition to be doggish. Another dog is unwilling to lead her owner: a very good, cooperative heeler on leash, she won't pull ahead. The pair stands transfixed and bemused. A third dog will take weeks to learn that it's okay, here, to jump up on the table and grab the good-smelling food that he's found. This behavior would be called counter surfing in some circles — nosing for food left untended on the dining table or kitchen counter (depending on the height of the dog) — and is frowned upon by polite owners and their polite dogs.

Indeed, all of the well-trained and obedient dogs have something to overcome here. Obedience is "totally unnecessary," for nosework, one trainer suggests. Another: "in many cases, obedience is a hindrance" for tracking. If the dog is concerned about what his owner wants him to do at every moment — if his concerns are reduced to when to heel or when to sit — then he has lost some part of his dogness.

Happily, Finnegan has not been well trained. While he is the more civilized of my two dogs, highly responsive to me, after a fashion, he is allowed to be a dog with us, and to poke his nose in our bags — *Where have you been, and what have you brought me?* — in a way that turns out to be incredibly useful if you want your dog to find a bit of cheese hidden in a stranger's suitcase. Should that come up in your life.

The first stage of nosework is simply to introduce the dogs to the possibility that, when you walk into a space with your person, she will let you sniff wherever you want, and you may find Very Good Treats. Among our small group, this is a terrific and heady leap. After only a few rounds, some of the crated dogs are whimpering with excitement as they wait for their next turn.

White, lightweight boxes sit scattered in the room, ringed by movable gates. "They look like they could easily be destroyed," the owner of an extremely energetic pit mix worries. For Finn's first tour of the arena, Berger takes a few pieces of a treat from me, places them obviously in a box, jiggling the box for good measure and then dropping it on the ground. Finn goes right to the box. At this "discovery" I am to reward him copiously with further treats. While this seems a bit overkill now, the idea is that eventually he will be able to race ahead and find a hide on his own — and I will want him to *stay there* (eating the treats) until I can catch up with him.

The other dogs all have their turns and Finn's comes up again. This time the hide is a bit subtler. Finn takes off when I release him with "Go find it!" — but he doesn't find the hide straightaway. Instead, he's vacuuming the floor of the crumbs left by classes past. "The good thing is that he keeps searching," Berger spins it. I don't know that I would call it "searching," but I wait Finn out. After a very long floor-cleansing, Finn comes to the hide. We all applaud as though he's discovered the moon.

Over the next half dozen rounds Finn continues his wet-vacuum ways, but less and less. His gait changes: no longer casually ambling along; instead, he leaps toward the boxes. By the end of the session, Finn is standing up in his crate while he waits for his turn, watching the other dogs' attempts. En route home in the car, he reclines the seat, curls into a crescent, and falls asleep.

One week later, at the appointed hour, I leash up Finnegan and we head out to the car for the long commute to our class.

"We're going to nosework!" I propose — at which, Finn begins lunging down the street, a huge enthusiasm powering him forward. Although I know this is anatomically impossible, his eyes seem to widen into saucers. His ears are laid back against his head as though to streamline him for *faster-getting-there-now.*

This week, there is a new twist in the hides: the search space, as defined by accordion fences creating a nosework theater in the room, is larger. Indeed, from this day on, each class brings a slight change of the game. This is Berger's way of "expanding the world" of the dogs. One class, he aims to teach the dogs that "the world won't always be boxes" — and hides are concealed in baskets, egg cartons, buckets, under a sheet. Dogs learn prepositions: hides can be on, under, in.

On another day, the world is "tricky"; the scattered objects on the floor unexpectedly *never* hold hides and instead are distractors from where the hides really are: along the wall, or in plain sight. Some hides will be in corners, because, Berger says, "Corners are typically repugnant to dogs." I consider whether I have ever seen a dog heading for a corner, and cannot find an image to match that query. Clearly it is not part of typical hunting drive, say, to go for the corners.

On some days, the world the dogs will encounter has "problems": hides are in long plastic tubes with an opening too small for a head. They are in zipped suitcases. They are near the owners waiting with their charges, stoically not responding when dogs come around looking for a friendly scratch behind the ears. (Berger admits, "Some other instructors here wouldn't do this" — planting hides around the waiting owners — "because they think owners should be out-of-bounds. I think everything should be in bounds.")

One day, "the world is not just here" — Berger motions by the ground. "It's also up here," he says, fluffing a pillow of air at the height of his head. Hides are set on chairs, on boxes piled onto boxes. Later, they will be in cups nailed to walls or perched on windowsills.

All of the dogs are, initially, stymied by the elevated hides. Finn waltzes by one, then jerks his head back just after we passed. It takes an extra beat before, powered by his rapid motor of a wagging tail, he finally gets the hide from atop a tower of boxes. He then has no trouble finding a hide on a chair, sticking his head through the back to reach the hide on the seat. (Well, we have trained him for that with a young child in the house, I think.) Soon, the moment when the dogs suddenly sniff an elevated hide is visible to us all. Their nostrils turn obliquely as they pass, or, with the small-nosed, their heads cock, their bodies tense. Their gait changes from a walk to a prance with intent.

Eventually, the hidden scent will be not a treat but a "neutral" odor — one dogs are not predisposed to either like or dislike: the essential oils of birch, anise (the smell of licorice), or clove. Along the way, other incremental changes are introduced: searching outside, among vegetation; in hot weather, obliging smell-inhibiting panting; against the wind.

We think of bloodhounds, beagles, and their brethren as the "nose dogs," but what is especially satisfying about nosework is the range of dogs who can do it. Watch a nosework class or trial and maybe a quarter will be mixed-breed dogs — and of the others, plenty of Chihuahuas, schnauzers, and unlikely breeds appear. While law enforcement detection dogs swing strongly toward Labrador retrievers, German shepherds, and Belgian Malinois, none of these is in our class. Instead, the dogs in our first class range from a twenty-pound mixed-breed with long furry ears, to a mellow long-coated retriever, to a muscular pit mix who might top a hundred pounds. And yet each dog gets the game. Their working styles are as individual as their appearances. One Border collie habitually heads out in a measured but lively clockwise tour of the space. The long-eared dog is very laid-back about the whole endeavor, and alternates gentle sniffs with searching glances at his owner. The pit mix defines the antithesis, charging into the game with little owner consultation. While a few dogs will bravely stick their noses into a box or bag, this one works out a puzzle by stepping on it, grabbing it in his mouth, or pushing it out of the way. One day he actually ends up with a cone on his head, befitting his headstrong enthusiasm.

In my admittedly biased estimation, Finn is turning out to be a champion sniffer. He merrily launches out on every bout with a sprightly gait. He tours the periphery and works inward, poking his nose in crevices and nooks. He snuffles loudly and charmingly. I learn to see when Finn is telling me he thinks he has found something — the subtle change in his demeanor that is his "alert." His nosing lingers just a little longer; he paws the ground or air, referencing digging rather than actually digging. His searching is exhaustive, never giving up the hunt for that tiny piece of salmon, or waft of birch, that he knows is stashed somewhere.

This is miraculous to the human-nosed among us. On one go around the ring with Finn I try to sniff out the hide myself. It's birch; before Berger hides it I bring the container to my nose and smell a sharp mint note, which resolves into a eucalyptus chord. I watch where Berger hides it — between some folded chairs stacked against the wall — so I am, I'd think, nose-advantaged. As Finn and I approach the chairs, I direct my nostrils toward them and inhale. Finn, ahead of me, has his nose on the ground, leaving a slug trail of moisture on the mats. He's walked by the hide. I sniff again. While I am ineffectively washing my nose with dog-gym air, Finn has stopped walking and lifts his head high. He sniffs the wall where the chairs lean against it, then draws a line with his nose down, over, down . . . to right where the hide is. Before I can catch a whiff, he has wolfed down his reward.

After many weeks of nosework, it has become clear that the instruction is not just, or even primarily, for the dogs. It's for us. Sure, dogs need to learn the game. But they get it at once, and steadily proceed through each level of difficulty. It is the owners who need training — even to understand how good their dogs are. Owners do not get it. At this stage, by which point every dog in the bunch has displayed prodigious abilities to catch the scent of a minuscule salmon morsel in a closed box in the corner of the room, owners still walk out of the ring and say, *Do you think he was using his nose?* Or, *I think she just got it accidentally.* One afternoon Berger directs every owner to handle someone else's dog in a search; after a few bemused dog glances, each dog sets to his task with alacrity. And all of us get to see what our dogs actually do — without our assumptions getting in the way.

Every search by a blind yellow Lab in a later class is the ultimate proof: he cannot see visual landmarks, so he really airscents. Often he seems to sense an obstacle before hitting it: smelling it in front of him. He always finds the hides.

Not only do we have trouble seeing what the dogs are doing, we have trouble playing our role. Our role, incidentally, is mostly to stay out of the way: follow the dog, create the space for him to search, and step in only to reward him for finding the hide. Somehow, this is unintuitive. Even at the course's end Berger will be encouraging us to bother to *really praise* our dogs — by petting, with more treats, with praise — when they've succeeded. "Do them *all,*" he says. "It's a big deal" that they've found it. "The quality of your payment," Norwegian trainer Anne Lill Kvam says, is directly connected to "the quality of your results. If you pay the dog with tiny little pieces" of food, she says, "you tell them *okay*": that's not a bad thing they did. "If you give them big ones, you tell him *This is it!*"

Worse, we cheat. Even in this no-stakes game, we unintentionally bamboozle even ourselves. If I see where Berger has stashed the treat, I may unconsciously lead my dog . . . roughly right over to it. In the class, as we watch each other with our dogs, we witness this happening dozens of times. The leading is more subtle than calculated: for instance, if a dog is idling in one area, an owner may simply angle her body toward the part of the room that holds the hide. This cue is like shouting through a bullhorn to the dog: *GO OVER THERE! IT'S OVER THERE!!!* Or an owner dawdles when the dog runs toward what she knows to be a barren area, resulting in tension on the leash (screaming *STOP* to the dog). If the owner keeps walking as the dog slows, the dog takes that as a clue to keep moving along. Even the side of the dog's body on which an owner lets the leash fall can influence where the dog goes next. An owner's tension as the dog innocently saunters right by a hidden canister is carried down the leash to the dog. Just as the detection dog's handler's job is to be responsive to the dog, not leading of him, here, too, the owners are tasked with following, not steering the dog.

It is difficult to be entirely neutral. When I put on my stone-coldest face, try to follow Finn wherever he goes and give him no clues or pushes in the "right" direction, I still find my mind going to the hide. I wonder how my body is following it.

The cure for our unwitting cheating is simple, happily: blind hides, in which neither the dog nor the owner knows where the treat has been hidden. Once we graduated to blind hides the dogs' performance convinced even the most skeptical of owners. As one heads into the arena, one must truly observe the dog, follow him, and trust that wherever he goes, there's a good reason for it.

Here I find the one reliable place in my life where my own ignorance is a pleasure. I love not knowing: it is a great discovery to have my dog be the informant for me. On walks in both city and country I begin to view Finn's now-prodigious sniffing as telling me a story of what has happened on this path before we got here. An unearthed set of tracks shows where a flock of turkeys walked, then alighted; a small burrow turns out to be fur-lined, perhaps the feeding den of a coyote. In the city, I know if our neighbor and his dog have been outside by watching Finn's examination of the air around our stoop, which they pass en route to the park.

After a few months, all the dogs are changed. One dog who began as a barker, reactive to other dogs, mostly ignores them in search of a quarter-inch of cubed chicken. Others have learned to look to their owners to *tell* the people something, not just to ask them or to get instruction; or that it's okay to put their nose on a table or gently jump up on a wall. Another dog, a collie mix, is like a cartographer. She tours the perimeter and will exhaustively examine every square meter within. When she is done with her survey, she beelines back to the smelly spot. All the dogs begin to lick their noses, better to snag the scent out of the air, in preparation to search. Finn's nose has become nearly periscopic, air-scenting up above the fray of the odor sea and sighting the hide on the horizon. I soon learn that I can't hide the tiniest dog treat on a windowsill and expect Finnegan not to find it. He slows as he walks by, checks the air, snorts, then pivots on a dime, sniffing great gulps right up to the smelly source. I feel perhaps unduly proud of him.

My own vision is changed: through the dogs, I've begun to see how the air in an ordinary room is moving. "Scent corridors" along walls — invisible tunnels of air that hurry odors in their wake — lead sniffing dogs to a hide from well across the room. Vents in the ceiling push air down and to the sides, creating small vortices along the middle of the room. Baseboard heaters push scent up the wall with the heated air. Currents flow from under doors and wash the room with the scent from outside.

At home, we begin doing nosework games in the living room, and Upton, watching the game, quickly picks it up. Their styles are different — Finn is an assiduous searcher, exhaustively visiting the home landmarks (couch, rocking chair, baseboards, bookshelves); Upton simply wanders around, nose in the air, and seems to catch odors on the rise — but both find their targets.

I begin trying other nose-search games

with Finn, such as asking him to pick up something I've dropped on our walk: first a highly valued orange ball, then a glove that smelled strongly of dog treats, then my keys. One day he finds my datebook for me, nosing it out with a satisfied expression. I stare at him in wonder — not only that he could find it, but that he knew what to find. *How much more do you know, my friend?*

12: Smelloftheworld

> The act of smelling something, anything, is remarkably like the act of thinking itself.
> — Lewis Thomas, *Late Night Thoughts on Listening to Mahler's Ninth Symphony*

The great pleasure in having spent the last years thinking about smelling is that my world has changed color. It smells. Well, it has always smelled, just as the light bouncing amidst the reds and greens of our palette are the unnoticed ultraviolets and infrareds that the honeybee and pit snake attend to. But I had not bothered to open my mind to the smells.

What has *not* happened is that I have become an expert in smelling. Nor am I constantly ringing with Proustian experiences, inflating ephemeral bubbles of memories with each inhale. I lived for over forty years not attending particularly to the *smelloftheworld,* in e. e. cummings's phrasing; so now there are only a handful of memories to reawaken. But I am now smelling all over the place. I put my nose to things without fear and with interest. I smell the skunkiness of bitter coffee beans swirled into ice cream. The loamy smell of the park after rain.* I smell when the previous occupant of our building's elevator has opened a new glossy magazine as she rode home. I smell someone's sudden urge to clean his hands with a sanitizing wipe while on the bus.

A summer trip in the car is not complete without notes of gasoline, cut grass, honeysuckle, warmed vinyl, sunscreen, overheated-dog breath, and wet sandals, stirred by wind from the open windows or wafting up from the floorboards.

I smell a thunderstorm approaching on my last visit to Colorado, the home of my family during my childhood, when I come to help clean out the house after my father's death. The watery, fresh smell of sea air comes, I now know, from ozone carried down from higher altitudes on the winds of a storm.† It is also the odor of the city when

* geosmin: "a metabolic by-product of bacteria or blue-green algae."

† Its "freshness" led to the earlier erroneous claim that ozone was health-bringing.

I emerge after swimming, the receptors pinging *chlorine!* silent for a long enough moment for me to smell the world in its absence.

I smell gin on the man who sits next to me in 10C.

I smell the acrid, lingering piles of freshly turned, festering wood chips on the other side of the park.

I see two people with a dog; then a second later smell that dog's poo, which must've recently been deposited in a trash bin.

I smell the art room at kindergarten before seeing it.

I smell every book I open.

I smell the clove antiseptic paste that screams *dental office.*

On return from a trip, I smell the way our apartment smells to those who do not live in it. I see that it is not difficult to know there are dogs living in it.

I smell how long ago an office door was last opened.

My nose captures vernal smells: the incomplete and jagged smells of fall, caught in winter, and finally released from their hibernation in spring.

One summer day I retronasally smell a good peach — slurpilly juicy — and orthonasally take in scrubbed sidewalks, a fetid sewer puddle, a vinegar sharpness.

I smell blue cheese in the building; Magic Marker on the street; cucumber on the east side of the park, steeping tea on the west side.

I smell my friends (sorry, friends, I have smelled you). Each person in my world has a "smell-face," as my great colleague Dr. Oliver Sacks once described it, and it is neither terrific nor terrifying; it simply rings of them.

I smell the wet wool on my fingers after coming in from a winter storm; coffee on my fingers after handling a takeout cup.

When I wake up, I take a moment to smell if my son or husband — or both — are still in bed. When I return home, I smell who has returned home before me.

When I awaken from the rare nap, I try to smell the time of day.

I smell burnt rubber by the train, smoked mozzarella in the Bronx, warmed pine needles in the forest.

Perhaps I will revisit these smells in another forty years and be brought back to these days, when I lived with a young boy, a youngish man, two great dogs, and an accidental cat, and headed into my books nose-first.

I have now dreamed of smells — smells of

people I love; and unknown, unsmelled smells concocted by my unconscious brain. Awake, I can now call forth a smell imaginatively, in my mind's nose. If I consider the smell of a penny . . . there, I've got it, the familiar waft that is really the smell not of copper and zinc but of *copper and zinc handled by persons.* As I write the word *lavender* (*lavender . . . lavender . . .*) I can conjure up the soapiness of the dried plant, sachet-ready. I look at a photograph of my father's desk, and the slightly sharp and resinous smell of its inside, mixed with pencil shavings and loose tobacco, hits me, just as if I had opened the drawer itself.

Smells precede us and remain after we leave: our presence is thus extended in a place. Watching the dogs has allowed me to extend the dimensions of perception.

Ultimately what I have learned to do is simply to bother to attend to smells. This was enabled by making associations — with words and with images — to fix my mind on a smell and then to curl it into a slip of memory. The pictorial and verbal vocabulary I am collecting is what clears access not just to current perception, but to my being able to smell more next time I nose a rose. In some important way, I realize that this is the opposite of being a dog, though.

I am taking advantage of something that we have that the dog does not — the words that describe, and simply accompany my experience. So I must assume that I am not reproducing Finn's experience.

While smells now appear to me more *public* — they are out there to be detected by a nose — I am evermore appreciative of the privacy of smells. Odors may blow in from across the river or waft through an open window from outside, but the greatest majority of smells need to be nearly touched to be perceived. Smell is a private sense, reserved for those people and objects we literally bring close to us. By smelling what my dog does, it brings me closer to him.

Today — as I will do tomorrow, and the next tomorrow — I head out the door with both my dogs. I watch their idle sniffs, their exploratory sniffs, their communicative sniffs, their pushy sniffs. When my dogs poke their noses in the earth, I pause to let them. I feel a frisson of excitement — both from knowing some of what is happening, and from the realization that I will never really know what is happening.

I will never smell as a dog does. I accept it. It is dogs' difference I celebrate — and their ways of smelling — their very noses —

are different. Quiet distillers of a world that we have stood up from and forgotten.

NOTES AND SOURCES

Being a dog emerged from conversations, interviews, classes, and experiences with dogs and people who work with or think about smell. On the person side, this includes Jonathan Ball, George Berger, John Buechsenstein, Noah Charney, Brent Craven, Annemarie DeAngelo, Bob Dougherty, Charley Eiseman, Stuart Firestein, Simon Gadbois, Avery Gilbert, Leta Herman, Pat Kaynaroglu, Raymond Matts, Alana McGee, Kate Mclean, Cindy Otto, George Preti, Kristin Rosenbach, Leslie Vosshall, and Sam Wasser.

Many books on smell served as reference, inspiration, and merely good riffling. I recommend them all:

Ackerman, D. 1990. *A natural history of the senses.* New York: Vintage Books.

Doty, R. L., ed. 2003. *Handbook of olfaction*

and gustation, 2nd ed. New York: Marcel Dekker, Inc.

Drobnick, J., ed. 2006. *The smell culture reader.* Oxford: Berg.

Feigel, L., ed. 2006. *A Nosegay: A literary journey from the fragrant to the fetid.* London: Old Street Publishing.

Gerritsen, R., and R. Haak. 2015. *K9 Scent training: A manual for training your identification, tracking, and detection dog.* Canada: Brush Education.

Gilbert, A. 2008. *What the nose knows: The science of scent in everyday life.* New York: Crown Publishers.

Henshaw, V. 2013. *Urban smellscapes: Understanding and designing city smell environments.* London: Routledge.

Rezendes, P. 1999. *Tracking and the art of seeing: How to read animal tracks and sign.* New York: HarperCollins.

Rouby, C., B. Schaal, D. Dubois, R. Gervais, and A. Holley, eds. 2002. *Olfaction, taste, and cognition.* Cambridge: Cambridge University Press.

Additional sources for each chapter are listed below. (I also include additional notes here and there, for the extra-interested reader.)

2: Smeller

On Dogs' Sensitivity in Finding Banana, Butyric Acid, and Human Odors:

Walker, D. B., J. C. Walker, P. J. Cavnar, J. L. Taylor, D. H. Pickel, S. B. Hall, and J. C. Suarez. 2006. Naturalistic quantification of canine olfactory sensitivity. *Applied Animal Behaviour Science, 97,* 241–254.

Neuhaus, W., and S. R. Lindsay. 2000. *Handbook of applied dog behavior and training, Vol. 1: Adaptation and learning.* Ames, Iowa: Blackwell Publishing.

Sulimov, K. T., V. I. Starovoitov, T. F. Moiseeva, I. I. Poletaeva, and E. P. Zinkevich. 1995. Dogs distinguish by scent quantitatively different mixtures of three higher fatty acids. *Sensory Systems, 9,* 99–102.

On Following Scent Trails:

Sommerville, B., and M. Green. 1989. The sniffing detective. *New Scientist, 122,* 54–57.

Discovery Channel Show

MythBusters: "Dog Myths." Aired March 14, 2007.

Five Footsteps

Hepper, P. G., and D. L. Wells. 2005. How many footsteps do dogs need to determine the direction of an odour trail? *Chemical Senses, 30,* 291–298.

Scent-Identification Lineups:

Schoon, G. A. A. 1996. Scent identification lineups by dogs (*Canis familiaris*): Experimental design and forensic application. *Applied Animal Behaviour Science, 49,* 257–267.

"Signature Odor" of Anal Sac:

Bradshaw, J. 2011. *Dog sense: How the new science of dog behavior can make you a better friend to your pet.* New York: Basic Books.

Anal Sac Research:

Preti, G., E. L. Muetterties, J. M. Furman, J. J. Kennelly, and B. E. Johns. 1976. Volatile constituents of dog (*Canis familiaris*) and coyote (*Canis latrans*) anal sacs. *Journal of Chemical Ecology, 2,* 177–186.

Doty, R. L., and I. Dunbar. 1974. Attraction of beagles to conspecific urine, vaginal and anal sac secretion odors. *Physiology & Behavior, 12,* 825–833.

Smell of Fox:

Doty 2003.

Females Sniff Faces First:

Bradshaw 2011.

Apocrine Glands on Pads of Dogs' Feet:

Hepper, P., and D. Wells. 2015. Olfaction in the Order Carnivora: Family Canidae. In R. L. Doty, ed. 2015. *Handbook of olfaction and gustation,* 3rd ed (pp. 591–603). Hoboken, NJ: Wiley-Blackwell.

Other Animals' Urine-Marking:

SNOWSHOE HARES:

Rezendes 1999; Liebenberg, L. 1990. *A field guide to the animal tracks of Southern Africa.* South Africa: David Philip Publishers.

RHINO AND HIPPO:

Watson, L. 2000. *Jacobson's organ and the remarkable nature of smell.* New York: W. W. Norton & Company.

BUSH DOG:

Porton, I. 1983. Bush dog urine-marking: Its role in pair formation and maintenance. *Animal Behaviour, 31,* 1061–1069.

Mice Counter-Marking:

Rich, T. J., and J. L. Hurst. 1999. The competing countermarks hypothesis: Reliable assessment of competitive ability by potential mates. *Animal Behaviour, 58,* 1027–1037.

Ferkin, M. H., and A. A. Pierce. 2007. Perspectives on over-marking: Is it good to be on top? *Journal of Ethology, 25,* 107–116.

The "Law of Urination":

Yang, P. J., J. C. Pham, J. Choo, and D. L. Hu. 2014. Duration of urination does not change with body size. *Proceedings of the National Academy of Sciences, 111,* 11932–11937.

Domestic Dog Marking Behavior:

Berthoud, D. 2010. Communication through scents: Environmental factors affecting the urine marking behaviour of the domestic dog, *Canis familiaris,* kept as a pet. PhD thesis, Anglia Ruskin University.

Lisberg, A. E., and C. T. Snowdon. 2011. Effects of sex, social status and gonadectomy on countermarking by domestic dogs, *Canis familiaris. Animal Behaviour, 81,* 757–764.

Objects and Style of Scent-Rolling:

Gosling, L. M., and H. V. McKay. 1990. Scent-rubbing and status signalling by mammals. *Chemoecology, 1,* 92–95.

Koler-Matznick, J., I. Lehr Brisbin Jr., and M. Feinstein. 2005. An ethogram for the New Guinea Singing (Wild) Dog (*Canis hallstromi*). The New Guinea Singing Dog Conservation Society, U.S.A.

Theories of Scent-Rolling:

Drea, C. M., S. N. Vignieri, S. B. Cunningham, and S. E. Glickman. 2002. Responses to olfactory stimuli in spotted hyenas (*Crocuta crocuta*): Investigation of environmental odors and the function of rolling. *Journal of Comparative Psychology, 116,* 331–341.

McCormick, J. 1993. In praise of stinks. *The Lancet, 341,* 1126–1127.

Ryon, J., J. C. Fentress, F. H. Harrington, and S. Bragdon. 1986. Scent rubbing in wolves (*Canis lupus*): the effect of novelty. *Canadian Journal of Zoology, 64,* 573–577.

Soldiers Returning Home:

see e.g., https://www.youtube.com/watch?v=eZ6oS5dUT30

Mirror Mark Studies:

Gallup Jr., G. G. 1970. Chimpanzees: Self-recognition. *Science, 167,* 86–87; Plotnik, J. M., F. B. M. de Waal, and D. Reiss. 2006. Self-recognition in an Asian elephant. *Proceedings of the National Academy of Sciences, 103,* 17053–17057; Reiss, D., and L. Marino. 2001. Mirror self-recognition in the bottlenose dolphin: A case of cognitive convergence. *Proceedings of the National Academy of Sciences, 98,* 5937–5942.

Study of Dog Urine in Snow:

Bekoff, M. 2001. Observations of scent-marking and discriminating self from others by a domestic dog (*Canis familiaris*): Tales of displaced yellow snow. *Behavioural Processes, 55,* 75–79.

Smells in Wind:

Hull, J. M. 1990. *Touching the rock: An experience of blindness.* New York: Vintage Books.

Smell of Storm Approaching:

Ackerman 1990. From Ackerman's observation about cows' anticipatory behavior before a storm.

Right- and Left-Nostril Sniffing:

Siniscalchi, M., R. Sasso, A. M. Pepe, S. Dimatteo, G. Vallortigara, and A. Quaranta. 2011. Sniffing with the right nostril: Lateralization of response to odour stimuli by dogs. *Animal Behaviour, 82,* 399–404.

Self-Experimenters:

Fiks, A. P. 2003. *Self-experimenters: Sources for study.* Westport, CT: Praeger.

Weber's Experiments:

Mainland, J., and N. Sobel. 2006. The sniff is part of the olfactory percept. *Chemical Senses, 31,* 181–196.

New Guinea Singing Dog Sniff:

Koler-Matznick et al. 2005.

Airflow of Dog Sniff:

Settles, G. S., D. A. Kester, and L. J. Dodson-Dreibelbis. 2003. The external aerodynamics of canine olfaction. In F. G. Barth, J. A. C. Humphrey, and T. W. Secomb, eds. 2003. *Sensors and sensing in biology and engineering* (pp. 323–355). New York: SpringerWein.

Sir Satan:

Steen, J. B., I. Mohus, T. Kvesetberg, and L. Walløe. 1996. Olfaction in bird dogs during hunting. *Acta Physiologica Scandinavica, 157,* 115–119.

Size of Olfactory Epithelium:

Gerritsen and Haak 2015.

Number of Odors Detectable:

Firestein, S. 2001. How the olfactory system makes sense of scents. *Nature, 413,* 211–218.

Lock and Key Theory of Receptors:

This model is not without its detractors. Nineteenth-century models suggested that smell perception resulted from odor "undulations" or waves, as sound and vision perception do; in the twentieth century, Luca Turin elaborated this model and posited a vibrational theory of odor reception. (see Block et al. 2014. Implausibility of the vibrational theory of olfaction. *Proceedings of the National Academy of Sciences, 112,* e2766–e2774.) For the most recent response to this idea, see Vosshall, L. B. 2015. Laying a controversial smell theory to rest. *Proceedings of the National Academy of Sciences, 112,* 6525–6526.

"Key in Pocket" Theory:

Shepherd, G. M. 2012. *Neurogastronomy: How the brain creates flavor and why it matters.* New York: Columbia University Press.

On Dog Genome, Proportion Committed to Olfactory Receptor Genes:

Lindblad-Toh, K., C. M. Wade, and T. S. Mikkelsen, et al. 2005. Genome sequence, comparative analysis and haplotype structure of the domestic dog. *Nature, 438,* 803–819.

Ostrander, E. 2007. Genetics and the shape of dogs. *American Scientist, 95,* 406–413.

Genetic Differences in Performance on Olfactory Tests:

Hepper and Wells 2015.

"Brains Did the Smelling, and the Nose Was Just the Conduit":

such as Galen, see Totelin, L. 2015. Smell as sign and cure in ancient medicine. In M. Bradley, ed. 2015. *Smell and the ancient senses* (pp. 17–29). New York: Routledge.

Nose as Brain's Fan:
as said by David Chudnovsky. In Preston, R. March 2, 1992. "The Mountains of Pi." *The New Yorker.*

On Cajal's Work:
Figueres-Oñate, M., Y. Gutiérrez, and L. López-Mascaraque. 2014. Unraveling Cajal's view of the olfactory system. *Frontiers in Neuroanatomy, 8,* 55.

Size of Olfactory Lobe in Dog and Human:
Laska, M., and L. T. Hernandez Salazar. 2015. Olfaction in nonhuman primates. In R. L. Doty, ed. 2003. *Handbook of olfaction and gustation,* 2nd ed. (pp. 605–621).

Topographical Layers of Olfactory Bulb:
Bakker, J. 2013. Olfaction. In D. W. Pfaff, ed., *Neuroscience in the 21st century: From basic to clinical* (pp. 815–837). New York: Springer-Verlag.

"Odor of Decayed Worms":
Adrian, E. D. 1942. Olfactory reactions in the brain of the hedgehog. *Journal of Physiology, 100,* 459–473.

Cheddar-Based Experiments:
Shepherd 2012.

Research Exposing Rats to Predators' Glandular Secretions:
Zibrowski, E. M., and C. H. Vanderwolf. 1997. Oscillatory fast wave activity in the rat pyriform cortex: relations to olfaction and behavior. *Brain Research, 766,* 39–49.
For those interested in this kind of thing, the weasel odorant was 2-propylthietane; the red fox, trimethyl thiazoline.

Brain Response to "Owner Smell":
Berns, G. S., A. M. Brooks, and M. Spivak. 2014. Scent of the familiar: An fMRI study of canine brain responses to familiar and unfamiliar human and dog odors. *Behavioural Processes, 110,* 37–46.

Flehmen Response:
Sommerville, B. A., and D. M. Broom. 1998. Olfactory awareness. *Applied Animal Behaviour Science, 57,* 269–286.
Galizia, C. G., and P-M. Lledo. 2013. Olfaction. In C. G. Galizia, and P-M. Lledo, eds. *Neurosciences: From molecule to behavior: A university textbook* (pp. 253–284). Heidelberg: Springer-Verlag.

Pheromones and VNO:

Karlson, P., and M. Lüscher. 1959. "Pheromones": A new term for a class of biologically active substances. *Nature, 183,* 55–56.

Wyatt, T. D. 2014. *Pheromones and animal behavior: Chemical signals and signatures,* 2nd ed. Cambridge: Cambridge University Press.

Androstenone:

Sell, C. S. 2014. *Chemistry and the sense of smell.* Hoboken, NJ: John Wiley & Sons, Inc.

Bombykol:

Barnard, C. 2003. *Animal behaviour: Mechanism, development, function and evolution.* Canada: Pearson Education.

VNO:

Barrios, A. W., P. Sánchez-Quinteiro, and I. Salazar. 2014. Dog and mouse: Toward a balanced view of the mammalian olfactory system. *Frontiers in Neuroanatomy, 8,* 106.

Use of the Tail-Wag as Scent-Spreader:

Hickman, G. C. 1979. The mammalian tail: A review of functions. *Mammal Review, 9,*

143–157.
Lewin, V., and J. G. Stelfox. 1967. Functional anatomy of the tail and associated behaviour in woodland caribou. *Canadian Field-Naturalist, 81,* 63–66.

Some Air-Scenting Dogs Clear Their Noses by Lifting Their Heads:
Gerritsen and Haak 2015.

4: Walking While Smelling

"Smelliest [Square] Blocks":
http://sensorymaps.com/portfolio/new-yorks-smelliest-block/

Porteous's Smellscapes:
Porteous, J. D. 1990. *Landscapes of the mind: Worlds of sense and metaphor.* Toronto: University of Toronto Press.
Drobnick 2006.

In Ancient Times, Temples Mixed Milk and Saffron into Their Plaster:
Classen, C., D. Howes, and A. Synnott. 1994. *Aroma: The cultural history of smell.* London: Routledge.

Mosques Were Built with Musk and Rose Water Worked into Their Mortar:
Ackerman 1990.

Smells of the Week:
Porteous 1990.

Netherlands Pedestrian Plazas:
from Kate McLean.

"One Hundred Sites of Good Fragrance":
Japanese Ministry of the Environment, https://www.env.go.jp/air/kaori/. Retrieved July 2015.

Smell of Old Books:
Strlić, M., J. Thomas, T. Trafela, L. Cséfalvayová, I. Kralj Cigić, J. Kolar, and M. Cassar. 2009. Material degradomics: On the smell of old books. *Analytical Chemistry, 81,* 8617–8622.

Buchbauer, G., L. Jirovetz, M. Wasicky, and A. Nikiforov. 1995. On the odor of old books. *Journal of Pulp and Paper Science, 21,* 398–400.

Manhattan Grid Design:
Drobnick 2006, p. 114.

Henshaw 2013.

Smells of Paris and London:

Keate, G. 1802. *Sketches from nature: Taken, and coloured, in a journey to Margate (1802).* London: T. Hurst.

Margolies, E. 2006. Vagueness gridlocked: A map of the smells of New York. In J. Drobnick, ed. (pp. 107–117).

Henshaw 2013.

Reinarz, J. 2014. *Past scents: Historical perspectives on smell.* IL: University of Illinois Press.

"Deodorization" Projects:

Drobnick 2006.

Homogenization of City Smells:

Drobnick 2006.

Henshaw Walks to Capture Smellscapes:

Henshaw 2013.

Honey Smell of Paris:

Stromberg, J. June 7, 2013. "Mapping the smells of New York, Amsterdam and Paris, block by block." *Smithsonian* magazine.

Correspondences Between Color Names and Odors:

Gilbert, A. N., R. Martin, and S. E. Kemp. 1996. Cross-modal correspondence between vision and olfaction: The color of smells. *The American Journal of Psychology, 109,* 335–351.

Maple Syrup Smell:

DePalma, A. October 29, 2005. "Good smell vanishes, but it leaves air of mystery." *New York Times.*

Lindeman, S. June 14, 2010. "The mystery of the maple syrup smell." *The Atlantic.*

Blindfolded Undergrads Navigating by Smell:

Jacobs, L. F., J. Arter, A. Cook, and F. J. Sulloway. 2015. Olfactory orientation and navigation in humans. *PLOS ONE, 10,* e0129387.

Sailors Use Smell in Navigation:

Beck, H. 1973. *Folklore and the sea.* Middletown, CT: Wesleyan University Press.

Homing Pigeon:

Jacobs, L. F. 2012. From chemotaxis to the cognitive map: The function of olfaction. *Proceedings of the National Academy of Sciences, 109,* 10693–10700.

Dogs in the First World War:
Richardson, E. H. 1920. *British war dogs: Their training and psychology* (pp. 171–172). London: Skeffington & Son Ltd.

5: Plain as the Nose on Your Face

Almost All Living Creatures Smell:
Doty 2003.

Sense People Are Most Willing to Lose:
Drobnick 2006 — though people who come to suffer from anosmia tend to disagree.

Asian Elephant Olfaction:
Rizvanovic, A., M. Amundin, and M. Laska. 2013. Olfactory discrimination ability of Asian elephants (*Elephas maximus*) for structurally related odorants. *Chemical Senses, 38,* 107–118.

"Smell of Metal":
Glindemann, D., A. Dietrich, H-J. Staerk, and P. Kuschk. 2006. The two odors of iron when touched or pickled: (Skin) carbonyl compounds and organophosphines. *Angewandte Chemie International Edition, 45,* 7006–7009.

"Principal Axis of Human Odor Perception . . . Remains Odor Pleasantness":
Yeshurun, Y., and N. Sobel. 2010. An odor is not worth a thousand words: From multidimensional odors to unidimensional odor objects. *Annual Review of Psychology, 61*, 219–241.

"The Greatest Poets in the World":
Woolf, V. 1933. *Flush: A biography.* New York: Harcourt Brace Jovanovich.

Freud: Sublimation of Smell:
Freud, S. 1978. In Le Guérer, A. 2002. Olfaction and cognition: A philosophical and psychoanalytic view. In Rouby et al. 2002 (pp. 3–15).

Odorous Is Odious:
Drobnick 2006, p. 14.

Smell of the Security Blanket:
Wyatt 2014.

They Don't Know That "Skunk" Is a Bad Smell:
originally posited by Freud (1929/1961). *Civilization and its discontents,* trans. J. Strachey. New York: W. W. Norton &

Company; later confirmed (with alterations) experimentally.

Kinds of Animal Noses:

MOLLUSKS:

Ache, B. W., and J. M. Young. 2005. Olfaction: Diverse species, conserved principles. *Neuron, 48,* 417–430.

NEMATODES:

Hart, A. C., and M. Y. Chao. 2010. From odors to behaviors in *Caenorhabditis elegans.* In A. Menini, ed. *The Neurobiology of Olfaction.* Boca Raton, FL: CRC Press/Taylor & Francis.

MOLES:

Catania, K. C. 1999. A nose that looks like a hand and acts like an eye: The unusual mechanosensory system of the star-nosed mole. *Journal of Comparative Physiology A, 185,* 367–372.

WATER SHREW:

Catania, K. C. 2006. Underwater 'sniffing' by semi-aquatic mammals. *Nature, 444,* 1024–1025.

DEET as Molecular Confusant:

Pellegrino, M., N. Steinbach, M. C. Stensmyr, B. S. Hansson, and L. B. Vosshall. 2011. A natural polymorphism alters odour and DEET sensitivity in an insect odorant receptor. *Nature, 478,* 511–514.

Categories of Noses:

Laska and Hernandez Salazar 2015.

"Has No Exotic Uses":

Asimov, I. 1963. *The human body: Its structure and operation.* Cambridge, MA: The Riverside Press.

"Gale-Force Speeds":

Clerico, D. M., W. C. To, and D. C. Lanza. 2003. Anatomy of the human nasal passages. In R. L. Doty, ed. 2003. *Handbook of olfaction and gustation,* 2nd ed (pp. 3–31). New York: Marcel Dekker, Inc.

Human Pheromones:

Wyatt 2014.

Green Cherry Juice Tastes Like Lime:

Sela, L., and N. Sobel. 2010. Human olfaction: A constant state of change-blindness. *Experimental Brain Research, 205,* 13–29.

"Baby Smells an Odor, Mother Says Nothing":

Roach, M. 2013. *Gulp: Adventures on the alimentary canal.* New York: W. W. Norton & Company.

Eighteen-Inch Buffer of Personal Space:

Hediger, H. 1950. *Wild animals in captivity.* London: Butterworth.

Ability to Smell Banana Odor:

Laska, M., A. Seibt, and A. Weber. 2000. 'Microsmatic' primates revisited: Olfactory sensitivity in the squirrel monkey. *Chemical Senses, 25,* 47–53.

Some Other Animals Smell Carbon Dioxide:

Mice and rats smell CO_2: Jones, W. 2013. Olfactory carbon dioxide detection by insects and other animals. *Molecules and Cells, 35,* 87–92.

Retronasal Olfaction:

Shepherd 2012.

Retronasal Olfaction in Dogs:

Craven, B. A., E. G. Paterson, and G. S. Settles. 2010. The fluid dynamics of canine olfaction: Unique nasal airflow patterns as an explanation of macrosmia. *Journal of the Royal Society Interface, 7,* 933–943.

Selective Anosmia Can Be Inherited; Genetic Thresholds to Detection:

Zhang, X., and S. Firestein. 2007. Nose thyself: Individuality in the human olfactory genome. *Genome Biology, 8,* 230.

6: My Dog Made Me Smell It

Inborn Skill at Smelling:

Porter, R. H., J. M. Cernoch, and F. J. McLaughlin. 1983. Maternal recognition of neonates through olfactory cues. *Physiology & Behavior, 30,* 151–154.

Schaal, B., L. Marlier, and R. Soussignan. 1995. Responsiveness to the odour of amniotic fluid in the human neonate. *Biology of the Neonate, 67,* 397–406.

Mallet, P., and B. Schaal. 1998. Rating and recognition of peers' personal odors by 9-year-old children: An exploratory study. *Journal of General Psychology, 125,* 47–64.

In addition to citations, some of this comes from Sela and Sobel 2010. Also Gilbert 2008; Gerritsen and Haak 2015; Wyatt 2014.

Nonhuman Animal Scent Recognition:

PAPER WASP AND BELDING'S GROUND SQUIRREL:

Alcock, J. *Animal Behavior.*

Spotted Hyena:

Emery, N. et al., eds. *Social intelligence: From brain to culture.*

Owner's Rating of Smell of Dog's Blanket:

Wells, D. L., and P. G. Hepper. 2000. The discrimination of dog odours by humans. *Perception, 29,* 111–115.

Smell of Lab Mice:

Gilbert, A. N., K. Yamazaki, G. K. Beauchamp, and L. Thomas. 1986. Olfactory discrimination of mouse strains (*Mus musculus*) and major histocompatibility types by humans (*Homo sapiens*). *Journal of Comparative Psychology, 100,* 262–265.

"You Just Smell the Books":

Feynman, R. P. 1985. *"Surely you're joking, Mr. Feynman!"* (pp. 105–106). New York: W. W. Norton & Company. Found via Gilbert 2008.

Bipedalism Demoting Smells:
E.g., Sigmund Freud (1929/1961) and Stuart Firestein (interview), among others.

Demotion of Smell with Promotion of Vision:
as laid out in Wyatt 2014 and Shepherd 2012.

James on Sensory Component of Practice:
James, W. 1890. *The Principles of Psychology,* vol. 1. New York: Henry Holt & Co.

Experiment on Smell Learning:
Li, W., J. D. Howard, T. B. Parrish, and J. A. Gottfried. 2008. Aversive learning enhances perceptual and cortical discrimination of indiscriminable odor cues. *Science, 319,* 1842–1845.

Chocolate Trail Following:
Porter, J., B. Craven, R. M. Khan, S-J. Chang, I. Kang, B. Judkewitz, J. Volpe, G. Settles, and N. Sobel. 2007. Mechanisms of scent-tracking in humans. *Nature Neuroscience, 10,* 27–29.

"The Lost Muscles of the Nose":

This title comes from a nearly selfsame paper in the journal *Aesthetic Plastic Surgery.*

Levator Labii Superioris:

Standring, S. 2015. *Gray's anatomy: The anatomical basis of clinical practice,* 41st ed. New York: Elsevier.

5 to 10 Percent of Air Makes It to Olfactory Epithelium:

Roach 2013.

Olfaction Is an Active Process:

Mainland, J., and N. Sobel. 2006. The sniff is part of the olfactory percept. *Chemical Senses, 31,* 181–196.

Sniff Vigor, Volume, Flow, and Value:

Mainland and Sobel 2006.

Even in Sleep, Our Brain Registers Smells:

Arzi, A., L. Shedlesky, M. Ben-Shaul, K. Nasser, A. Oksenberg, I. S. Hairston, and N. Sobel. 2012. Humans can learn new information during sleep. *Nature Neuroscience, 15,* 1460–1465.

Two Sniffs Are Better Than One:
Joel Mainland, personal communication. May 13, 2015.

Right- and Left-Nostril Differences:
Herz, R. S., C. McCall, and L. Cahill. 1999. Hemispheric lateralization in the processing of odor pleasantness versus odor names. *Chemical Senses, 24,* 691–695.

Discriminate Unknown, New Odors When Using the Right Nostril:
Savic, I., and H. Berglund. 2000. Right-nostril dominance in discrimination of unfamiliar, but not familiar, odours. *Chemical Senses, 25,* 517–523.

Olfactometer:
Totelin 2015.

No Basic Words for Fundamental Smells:
Sperber, D. 1975. *Rethinking symbolism.* Cambridge: Cambridge University Press.

Malay Olfactory Language:
Wnuk, E., and A. Majid. 2014. Revisiting the limits of language: The odor lexicon of Maniq. *Cognition, 131,* 125–138.
Majid, A., and N. Burenhult. 2014. Odors are expressible in language, as long as you

speak the right language. *Cognition, 130,* 266–270.

Difficulty Naming Familiar Odors:
Yeshurun and Sobel 2010.

Trigeminal Nerve:
Shusterman, D. 2009. Qualitative effects in nasal trigeminal chemoreception. *Annals of the New York Academy of Sciences, 1170, International Symposium on Olfaction and Taste* (pp. 196–201).

7: Nose to Grindstone

Old DuPont Plant, Now Penn's WDC:
http://www.upenn.edu/pennnews/current/2012-09-13/features/penn's-south-bank-23-acres-pure-potential. Retrieved October 15, 2015. http://www.workshopoftheworld.com/south_phila/dupont.html. Retrieved October 15, 2015.

Hemingway on the Smell of Death:
Hemingway, E. 1940. *For whom the bell tolls.* New York: Charles Scribner's Sons.

"Saccharine Putrescence":
O'Rourke, P. J. 1988. *Holidays in hell.* New York: Atlantic Monthly Press.

Mortuary Technician Carla Valentine:
A Life in Scents podcast, http://bit.ly/1WrtaHY. Retrieved November 1, 2015.

8: Nose-Wise

Early Cancer-Finding Dogs:
Williams, H., and A. Pembroke. 1989. Sniffer dogs in the melanoma clinic? *The Lancet, 333,* 734.
Church, J., and H. Williams. 2001. Another sniffer dog for the clinic? *The Lancet, 358,* 930.
Welsh, J. S., D. Barton, and H. Ahuja. 2005. A case of breast cancer detected by a pet dog. *Community Oncology, 2,* 324–326.

Volatiles in Cancer:
Wells, D. L. 2012. Dogs as a diagnostic tool for ill health in humans. *Alternative Therapies in Health and Medicine, 18,* 12–17.

Bladder Cancer Study:
Willis, C. M., S. M. Church, C. M. Guest, W. A. Cook, N. McCarthy, A. J. Bransbury, M. R. T. Church, and J. C. T. Church. 2004. Olfactory detection of human bladder cancer by dogs: Proof of principle study. *British Medical Journal, 329,* 712–714.

Smells in Urine:

Shirasu, M., and K. Touhara. 2011. The scent of disease: Volatile organic compounds of the human body related to disease and disorder. *Journal of Biochemistry, 150,* 257–266.

Prostate Cancer Study:

Cornu, J. N., G. Cancel-Tassin, V. Ondet, C. Girardet, and O. Cussenot. 2011. Olfactory detection of prostate cancer by dogs sniffing urine: A step forward in early diagnosis. *European Urology, 59,* 197–201.

Melanoma Detection Study:

Pickel, D., G. P. Manucy, D. B. Walker, S. B. Hall, and J. C. Walker. 2004. Evidence for canine olfactory detection of melanoma. *Applied Animal Behaviour Science, 89,* 107–116.

Odors of Breath:

Phillips, M., J. Herrera, S. Krishnan, M. Zain, J. Greenberg, and R. N. Cataneo. 1999. Variation in volatile organic compounds in the breath of normal humans. *Journal of Chromatography B, 729,* 75–88.

Lung Cancer Study:

McCulloch, M., T. Jezierski, M. Broffman, A. Hubbard, K. Turner, and T. Janecki. 2006. Diagnostic accuracy of canine scent detection in early- and late-stage lung and breast cancers. *Integrative Cancer Therapies, 5,* 30–39.

GC as Generating an "Arpeggio":

Gilbert 2008.

Peaks of GC:

Shepherd 2012.
Gilbert 2008.

"Fully Biocompatible and Patient Friendly Alarm System":

Chen, M., M. Daly, N. Williams, S. Williams, C. Williams, and G. Williams. 2002. Non-invasive detection of hypoglycaemia using a novel, fully biocompatible and patient friendly alarm system. *British Medical Journal, 321,* 1565–1566.

Diabetic-Alert Dogs:

Rooney, N. J., S. Morant, and C. Guest. 2013. Investigation into the value of trained glycaemia alert dogs to clients with type I diabetes. *PLOS ONE, 8,* e69921.

Hippocrates "Open Nose"

Le Guérer 2002.

"As Water Changes to Air . . .":

Plato, via Totelin 2015.

Early Uterine Theory; Pliny Cure for Smelly Armpits:

Totelin 2015, p. 27.

Linnaeus:

Linnaeus. 1764. Odores medicamentorum. *Amoenitates Academicae,* vol. 3 (pp. 183–201). Stockholm: Lars Salvius.

NB: Gilbert 2008 made this observation first.

Medicinal Effect of Fragrant and Repulsive Plants:

Schiller F. 1997. A memoir of olfaction. *Journal of the History of the Neurosciences, 6,* 133–146.

Galen:

Totelin 2015.

Helpful Sixteenth-century Instructional Manual:

"Remèdes, Préservatifs et Curatifs de Peste," 1562. In Feigel 2006.

Smell of Breath:

Kwak, J., and G. Preti. 2011. Volatile disease biomarkers in breath: A critique. *Current Pharmaceutical Biotechnology, 12,* 1067–1074.

Smell of Diseases, Psychiatric Problems, and Toxins:

Watson 2000.

Orient, J. M., ed. 2010. *Sapira's art and science of bedside diagnosis,* 4th ed. Philadelphia: Lippincott Williams & Wilkins.

Kenny, J. C. 1989. The valuing, educational preparation and diagnostic use of the olfactory sense in nursing practice. Dissertation, Adelphi University.

Use of Odors in Western Medicine:

E.g., footnote in L. Goldfrank, R. Weisman, and N. Flomenbaum. 1982. Teaching the recognition of odors. *Annals of Emergency Medicine, 11,* 22.

"The Characteristic Foul Odor of the Sputum Suggests Anaerobic Involvement":

Chung, G., and M. B. Goetz. 2000. Anaerobic infections of the lung. *Current Infectious Disease Reports, 2,* 238–244.

Battery of Vials for Sniff-Training:
Orient 2010.

A "Ten Test Tube Sniffing Bar":
Kenny 1989.

TCM's Method:
Liu, Z., and L. Liu eds. 2010. *Essentials of Chinese medicine,* vol. 1. London: Springer-Verlag.

Five Elements/Phases:
Unschuld, P. U. 1985. *Medicine in China: A history of ideas.* Berkeley: University of California Press.

Alcoholic, Corn Farmer, Coal Miners:
Behrman, A. D., and S. Goertemoeller. 2009. What is that smell? *Journal of Emergency Nursing, 35,* 263–264.

Conditions That Produce Distinctive Odors:
Doty 2001; Wyatt 2014; Kenny 1989.

Woman Smelling Parkinson's:
http://www.scientificamerican.com/article/one-woman-s-ability-to-sniff-out-parkinson-s-offers-hope-to-sufferers/. Retrieved November 22, 2015.

Sense of Smell in Early Parkinson's:

Doty, R. L., S. M. Bromley, and M. B. Stern. 1995. Olfactory testing as an aid in the diagnosis of Parkinson's disease: Development of optimal discrimination criteria. *Neurodegeneration, 4,* 93–97.

Scratch-and-Sniff:

Doty 2009; Wyatt 2014.

"My Genius Resides in My Nostrils":

Nietzsche, F. 1911/2004. *Ecce Homo* (p. 132). A. M. Ludovici (transl). Mineola, N.Y.: Dover.

Nose-Wise:

Oxford English Dictionary, OED online. Retrieved March 2015.

9: STINK-WAVES

Roald Dahl:

The Witches (p. 24): 1983. London: Puffin; "Jack and the Beanstalk," from *Revolting rhymes.* 1982. New York: Alfred A. Knopf.

Ribbon Snakes:

Gadbois, S., and C. Reeve. 2014. Canine olfaction: Scent, sign, and situation. In A. Horowitz, ed. *Domestic dog cognition and behavior: The scientific study of* Canis familiaris. Heidelberg: Springer-Verlag.

Environmental Contaminants:

Arner, L. D., G. R. Johnson, and H. S. Skovronek. 1986. Delineating toxic areas by canine olfaction. *Journal of Hazardous Materials, 13,* 375–381.

Sui Generis:

United States v. Place (1983).

Pliny:

1855. *The natural history of Pliny,* vol. 2. (Book VII, p. 314). Bostock, J., and H. J. Riley (transl). London: Henry G. Bohn.

Smell of Person on Pipe Bomb:

Curran, A. M., P. A. Prada, and K. G. Furton. 2010. Canine human scent identifications with post-blast debris collected from improvised explosive devices. *Forensic Science International, 199,* 103–108.

Detection of Cadavers in Water:

Killam, E. W. 1990. *The detection of human remains.* Springfield, IL: Charles C. Thomas.

Warren, C. 2013. *What the dog knows* (p.

211). New York: Touchstone.

Avalanche-Rescue Dogs:
Killam 1990.

Longevity of Human Smell on Objects:
Curran, Prada, Furton 2010.
Syrotuck, W. G. 1972. *Scent and the scenting dog* (p. 106). Mechanicsburg, PA: Barkleigh Productions.

Human "Scent Samples":
Lesniak, A., M. Walczak, T. Jezierski, M. Sacharczuk, M. Gawkowski, and K. Jaszczak. 2008. Canine olfactory receptor gene polymorphism and its relation to odor detection performance by sniffer dogs. *Journal of Heredity, 99,* 518–527.
Curran, A. M., S. I. Rabin, and K. G. Furton. 2005. Analysis of the uniqueness and persistence of human scent. *Forensic Science Communications, 7,* 2.

Skin Cell Sloughing:
Allen T., and G. Cowling. 2011. *The cell: A very short introduction* (p. 10). Oxford: Oxford University Press.

Sweat Generated in Sitting or Exercising:
Medeiros, D. M., and R. E. C. Wildman. 2012. *Advanced human nutrition,* 2nd ed. Sudbury, MA: Jones & Bartlett Learning.
Watson 2000.

Olf:
coined by Fanger, P. O. 1988. Perceived quality of indoor and ambient air. *Proceedings of the Indoor Ambient Air Quality Conference,* London, 365–376. In McCormick 1993.

Components of Sweat:
Curran, A. M., S. I. Rabin, P. A. Prada, and K. G. Furton. 2005. Comparison of the volatile organic compounds present in human odor using SPME-GC/MS. *Journal of Chemical Ecology, 31,* 1607–1619.

Clues in a Footprint:
Gerritsen and Haak 2015.
Wright, R. H. 1982. *The sense of smell.* Boca Raton, FL: CRC Press.

Wet Shoe Experiment:
Gerritsen and Haak 2015.

Sweat from Soles of Feet:
Syrotuck 1972.

"Dopamine Breeds":
Gadbois and Reeve 2014.

Norway Sniffing Study:
Thesen, A., J. B., Steen, and K. B. Døving. 1993. Behaviour of dogs during olfactory tracking. *Journal of Experimental Biology, 180,* 247–251.

"Professional Poop Chasers":
Wasser, S. K. 2008. Lucky dogs. *Natural History, 117,* 48–53.

Yellow Baboons:
Wasser, S. K., ed. 1983. *Social behavior of female vertebrates.* New York: Academic Press.

Effect of Zebra Finch Leg Bands:
Burley, N. 1988. Wild zebra finches have band-colour preferences. *Animal Behaviour, 36,* 1235–1237.

Effect of Human and Wolf Activity on Caribou Populations:
Wasser, S. K., J. L. Keim, M. L., Taper, and S. R., Lele. 2011. The influences of wolf predation, habitat loss, and human activity on caribou and moose in the Alberta

oil sands. *Frontiers in Ecology and the Environment, 9,* 546–551.

Dogs with Excessive Energy:
Wasser 2008.

Tucker's Orca-Scat Detection:
Ayres, K. L., R. K. Booth, J. A. Hempelmann, K. L. Koski, C. K. Emmons, R. W. Baird, K. Balcomb-Bartok, M. B. Hanson, M. J. Ford, and S. K. Wasser. 2012. Distinguishing the impacts of inadequate prey and vessel traffic on an endangered killer whale (*Orcinus orca*) population. *PLOS ONE, 7,* e36842.

Scent at Bus Stop:
Watson 2000, p. 72.

Armpit Sniffing of Kanum-Irebe:
Eibl-Eibesfeldt, I. 1971. *Love and hate: The natural history of behavior patterns* (p. 191). New York: Holt, Rinehart and Winston.

Schultze-Westrum 1968, as cited in Mykytowycz, R. 1985. Olfaction — A link with the past. *Journal of Human Evolution, 14,* 75–90.

Synchronization of Menstrual Cycles:
Stern, K., and M. K. McClintock. 1998. Regulation of ovulation by human pheromones. *Nature, 392,* 177–179. This result is still debated.

Cold, Calm Winter Days Are Good for Ground Tracking:
Gerritsen and Haak 2015.

"Novel in an Owl's Pellet":
Rezendes 1999.

Rains Can "Breathe New Life":
Gerritsen and Haak 2015.

Scent Posts:
Rezendes 1999.

Some Porcupine Facts:
http://www.nwf.org/news-and-magazines/national-wildlife/animals/archives/1994/prying-into-the-life-of-a-prickly-beast.aspx. Retrieved November 1, 2015.

Dog's Gait Is "Sloppy":
Rezendes 1999.

"Smelly" Kelly:

Jones, P. 1978. *Under the city streets: A history of subterranean New York.* New York: Holt, Rinehart and Winston.

The MTA now uses electronic "sniffers" that analyze air samples, alas. (Neuman, W. October 3, 2006. "M.T.A. to Upgrade Chemical-Detection System." *New York Times.*)

10: Civet Cats and Wet Dogs

Truffle Biology:

Rubini, A., C. Riccioni, S. Arcioni, and F. Paolocci. 2007. Troubles with truffles: Unveiling more of their biology. *New Phytologist, 174,* 256–259.

Kunzig, R. 2000. The biology of . . . truffles. Expensive and delectable, truffles are one crop modern agriculture can't tame. *Discover.*

"Have the Local Experts . . . Been Creating Little Digs in the Earth?":

Trappe, M., F. Evans, and J. Trappe. 2007. *Field guide to North American truffles: Hunting, identifying, and enjoying the world's most prized fungi.* New York: Ten Speed Press.

Truffle Descriptors:
From Alana McGee and Trappe, Evans, and Trappe 2007.

Chesterton:
Chesterton, G. K. 1914. *The flying inn.* New York: John Lane Co.

Greeks and Romans:
Doty 2003.

"Extends [One's Body] into the Space Beyond Which It Should":
Tullett, W. A Life in scents podcast, http://bit.ly/1XMs36f

If a Woman "Seduced or Betrayed into Matrimony" Any Gentleman by Use of Scents:
Doty 2003.

"Rear End of an Asian Cat":
Turin, L. 2006. *The secret of scent: Adventures in perfume and the science of smell* (p. 90). New York: HarperCollins.

Chanel No. 5:
Sell, C. S., ed. 2006. *The chemistry of fragrances: From perfumer to consumer,* 2nd ed. Cambridge, UK: The Royal Soci-

ety of Chemistry.
Smelling notes from Raymond Matts.

Smell "the Hands That Picked" the Flower:

Feigel 2006, p. 73.

"Show Promise on Olfactory and Psychological Tests":

www.ifraorg.org.

Perfumers' Brains:

Plailly, J., C. Delon-Martin, and J-P. Royet. 2012. Experience induces functional reorganization in brain regions involved in odor imagery in perfumers. *Human Brain Mapping, 33,* 224–234.

Ambergris:

Rice, D. W. 2008. Ambergris. In W. F. Perrin, B. Würsig, and J. G. M. Thewissen, eds. *Encyclopedia of marine mammals,* 2nd ed (p. 28). San Diego, CA: Academic Press.

Montaigne's "Mustachios, Which Are Full":

Montaigne M. de. 1580. C. Cotton (transl). *Essays of Montaigne.* New York: Edwin C. Hill.

Tea Expert Who Knows Whether the Plant Grew Near a Plum Tree:

As invoked by Huysmans, J-K. 1884. *Against the grain.* In Feigel 2006.

Perfumery Methods:

Cinquième Sens, "Introduction to the techniques and language of perfumery."

Preserve the Headspace of Endangered Flowers:

The flowers are also memorialized in Kaiser's book, *Scent of the Vanishing Flora.*

"Verbalize Their Olfactory Experience":

Royet, J-P., J. Plailly, A-L. Saive, A. Veyrac, and C. Delon-Martin. 2013. The impact of expertise in olfaction. *Frontiers in Psychology, 4,* 928.

Verbal Landmarks to Invoke a Smell:

Gilbert, A. N., M. Crouch, and S. E. Kemp. 1998. Olfactory and visual mental imagery. *Journal of Mental Imagery, 22,* 137–146.

"Change Detection" System:

This is *adaptation,* introduced in chapter 6, My Dog Made Me Smell It. "Change detection" from Herz, R. 2007. *The scent of desire: Discovering our enigmatic sense of smell.* New York: William Morrow.

Advice of an Animal-Tracking Handbook:

Young, J., and T. Morgan. 2007. *Animal tracking basics.* Mechanicsburg, PA: Stockpole Books.

Jahai:

Majid and Burenhult 2014.

Olfactory Hallucinations:

Leopold, D. 2002. Distortion of olfactory perception: Diagnosis and treatment. *Chemical Senses, 27,* 611–615.

Noble Wine Wheel Aromas:

Noble, A. C., R. A. Arnold, J. Buechsenstein, E. J. Leach, J. O. Schmidt and P. M. Stern. 1987. Modification of a standardized system of wine aroma terminology. *American Journal of Enology and Viticulture, 38,* 143–146.

Noble, A. C. 2009. Using the wine aroma wheel. Downloaded August 27, 2015, from http://winearomawheel.com/Websites/aromawheel/Images/userguide_2010.pdf.

TCA Blocks Olfactory Receptors:
Takeuchi, H., H. Kato, and T. Kurahashi. 2013. 2,4,6-Trichloroanisole is a potent suppressor of olfactory signal transduction. *Proceedings of the National Academy of Sciences, 110,* 16235–16240.

11: Nosed Out

Smelloftheworld:
cummings, e. e. 1925. "and this day it was Spring."

Quantity-Smelling Study:
Horowitz, A., J. Hecht, and A. Dedrick. 2013. Smelling more or less: Investigating the olfactory experience of the domestic dog. *Learning and Motivation, 44,* 207–217.

Dogs Are Not Using Smell to Capacity:
See, e.g., Polgár, Z., Á. Miklósi, and M. Gácsi. 2015. Strategies used by pet dogs for solving olfaction-based problems at various distances. *PLOS ONE, 10,* e0131610.

Obedience Is "Totally Unnecessary" . . . "A Hindrance":
Turid Rugaas; Anne Lill Kvam.

"Lost Keys" Game:
Kvam, A. L. and T. Rugaas. 2012. *Nosework Search games* DVD.

12: SMELLOFTHEWORLD

Geosmin:
Yuhas, D. July 18, 2012. "Storm scents." *Scientific American.*

Ozone:
R. Matts, personal communication; Yuhas 2012. Etymology: Oxford English Dictionary.

Smell of a Penny:
Glindemann, Dietrich, Staerk, Kuschk 2006.

THANKS, AN INDEX OF

book

- advocacy of, *superagent Kris Dahl, Caroline Eisenmann*
- enactors of, *Nan Graham, Colin Harrison, Roz Lippel, Susan Moldow, and Scribner*
- making of
    - art, *Vegar Abelsnes, Jaya Miceli*
    - editorial, *Colin Harrison (my ideal reader)*
    - editorial, assistance of, *Sarah Goldberg*
    - marketing and publicity, *Katie Monaghan*
    - production editorial and design, *Mia Crowley-Hald, Erich Hobbing*
- space for writing, *New York Society Library*
- willing and useful interlocutors on, *Brian Boyd, Betsy Carter, Alison Curry, Holly Fairbank, Glen Finkel, Elizabeth Hardin, Damon Horowitz, Ogden Thelonious Horowitz Shea, Jay Horowitz, Daniel Hurewitz,*

Elizabeth Kadetsky, Maira Kalman, Sally Koslow, Aryn Kyle, Maria Popova, Douglas Repetto, Ammon Shea, Timea Szell, Andy Tuck, Jennifer Vanderbes, Carlin Wing, Mark Woods

dog nose

amateur, *every dog in New York City, Finnegan, Upton, volunteer subjects in my studies*

experts on, *Brent Craven, Simon Gadbois, Gary Settles*

instructor of, *George Berger*

model, *Finnegan*

professional, *Penn Working Dog Center dogs; truffle dogs Callie, Cash, Da Vinci, Lolo; scat-detection dogs; tracking dogs*

studies of, *Barnard College, John Herrold and the NYC Parks Department, members of the Dog Cognition Lab, owner participants*

trainers of, *Jonathan Ball, Annemarie DeAngelo, Bob Dougherty, Pat Kaynaroglu, Cindy Otto, and everyone at the WDC; Alana McGee and Kristin Rosenbach; Sam Wasser*

human nose

demonstrators of abilities of, *John Buechsenstein, Noah Charney, Charley Eiseman, Leta Herman, Ray Matts, Kate*

ABOUT THE AUTHOR

Alexandra Horowitz is the author of the #1 *New York Times* bestseller *Inside of a Dog: What Dogs See, Smell, and Know* (2009) and *On Looking: A Walker's Guide to the Art of Observation* (2013). She teaches at Barnard College, where she runs the Dog Cognition Lab. She lives with her family and two large, highly sniffy dogs in New York City.

The employees of Thorndike Press hope you have enjoyed this Large Print book. All our Thorndike, Wheeler, and Kennebec Large Print titles are designed for easy reading, and all our books are made to last. Other Thorndike Press Large Print books are available at your library, through selected bookstores, or directly from us.

For information about titles, please call:
(800) 223-1244

or visit our Web site at:
http://gale.cengage.com/thorndike

To share your comments, please write:
Publisher
Thorndike Press
10 Water St., Suite 310
Waterville, ME 04901